SpringerBriefs in Animal Sciences

For further volumes:
http://www.springer.com/series/10153

Vladimir V. Gouli · Svetlana Gouli
José A. P. Marcelino

Common Infectious Diseases of Insects in Culture

Diagnostic and Prophylactic Methods

 Springer

Dr. Vladimir V. Gouli
Entomology Research Laboratory
Department of Plant and Soil Science
University of Vermont
661 Spear Street, South Burlington
VT 05405-0105, USA
e-mail: vgouli@uvm.edu

Dr. Svetlana Gouli
Entomology Research Laboratory
Department of Plant and Soil Science
University of Vermont
661 Spear Street, South Burlington
VT 05405-0105, USA
e-mail: sgouli@uvm.edu

Dr. José A. P. Marcelino
Department of Biology
University of the Azores
Rua Mae de Deus
9500-321 Ponta Delgada
Azores, Portugal
e-mail: jmarcelino@uac.pt

ISSN 2211-7504
ISBN 978-94-007-1889-0
DOI 10.1007/978-94-007-1890-6
Springer Dordrecht Heidelberg London New York

e-ISSN 2211-7512
e-ISBN 978-94-007-1890-6

Cover design: eStudio Calamar, Berlin/Figueres

Printed on acid-free paper

Springer is part of Springer Science+Business Media (www.springer.com)

Preface

This handbook gives a thorough, comprehensible and copious illustrated description of the most common diseases in laboratory reared insect colonies, comprising Viruses (Baculoviridae, Reoviridae, Poxviridae, and Iridoviridae); Bacteria (Bacillaciae, Pseudomonadaceae and Enterobacteriaceae); Rickettsia (Rickettsiaceae); Protozoa (Amebiases, Gregarine and Coccidian); Fungi (Entomophthorales and Muscardine) and Microsporidia. Major aspects in insect disease detection and control, as well as prophylactic measures, are covered. The morphology of the pathogen, its external signs and symptoms in infected hosts, the pathomorphism of disease and the most common insect host for each pathogen group (Chap. 1) and for infection diseases in insect cultures, in general (Chap. 2) were described and depicted. In addition, efficient and accessible methods for practical diagnostic of disease were described in detail (Chap. 3), in a step by step protocol format, for an easy orientation and accurate learning. Prophylactics and control of infectious disease in laboratory reared insect colonies were also covered in detail (Chap. 4). Descriptions of the different species or families of microorganism agents of disease varied according to the current scientific knowledge of the organisms being described, its pathogenic importance and management potential.

The recommendations and protocols described here could be of critical importance for the establishment of axenic colonies of insects (from known or uncertain origins), not only for laboratories with mass rearing programs but for individual research programs including an insect rearing component. The initial phase of establishment of these colonies should imperatively include protocols to visually and microscopically assess the health of the individuals which will form the baseline for the future colony. Also, protocols for the periodic assessment of a colony health status should be incorporated into mass rearing activities in order to avoid cryptic infections in the colonies and possible spread of infection. While in advance stages of infection the only alternative is the eradication of the insect colony, in most cases an early detection can avoid this loss and the economic costs associated to it. Thus, it is necessary that all personnel working in insect rearing programs to be acquainted with fundamental knowledge on symptoms and signs of disease that may arise in early stages of infection, this way enhancing the health, reproductive performance and vigor of the colonies.

The present handbook offers the opportunity to laboratory personnel or individuals developing work in the field of entomology, which may not possess a robust background in invertebrate pathology, to acquire the necessary skills and knowledge in insect pathology in order to:

1. Provide aseptic conditions for mass rearing of insect colonies;
2. Assess the health condition of an insect colony at establishment and over time;
3. Identify the possible causal agent of disease through scientific protocols;
4. Determine the measures to take into action after identification of a disease.

Acknowledgments

We would like to acknowledge the support received from Zuzana Bernhart and Elisabete Machado, respectively Senior editor and publishing assistant of Plant Pathology and Entomology at Springer, The Netherlands. We are also grateful for the critical reviews of the manuscript by anonymous reviewers, as well as, several postdoctoral fellows, students and technicians which, through the years, helped to more accurately describe, prevent and incorporate into this handbook pathologies in insect reared colonies.

Contents

Chapter 1
Introduction

1.1 Handbook Objective

The purpose of the present handbook is to provide a comprehensive description of diseases found in laboratory reared insect colonies and to propose practical prophylactic and control measures for these pathologies. Information is supported with profuse original illustrations of signs and symptoms of disease.

There are numerous publications, released worldwide, that deal with insect diseases (see Tanada and Kaya 1992; Boucias and Pendland 1998; Solter and Becnel 2000; Hajek et al. 2004). Descriptive handbooks for the diagnostic of insect diseases have been published in the past (Poinar and Thomas 1984; Gouli and Ribina 1988; Lacey et al. 1997, 2007), however, these publications are intended for professional insect pathologists or for proficient training in invertebrate pathology. They are extensive and require a solid scientific background in insect pathology. Currently, most guides are web based, not comprehensive and solely addressing the most common diseases of a particular insect, or group of insects, reared in laboratory facilities, or found elsewhere. A comprehensive, practical and accessible handbook, for students and technicians working with insect reared colonies, is lacking. The current handbook aims to provide the basic background information on most insect diseases that can occur in laboratory or industrial insect populations. Although pathologies of many groups of entomopathogens still remain practically unexplored, extensive information associated with the pathologies of phytophagous insects allows for a broader understanding of the infection processes and general etiology in these animals and to extrapolate some of this knowledge to other insect groups (e.g., entomophagous species).

V. Gouli et al., *Common Infectious Diseases of Insects in Culture*,
SpringerBriefs in Animal Sciences, DOI: 10.1007/978-94-007-1890-6_1,
© Vladimir Gouli 2011

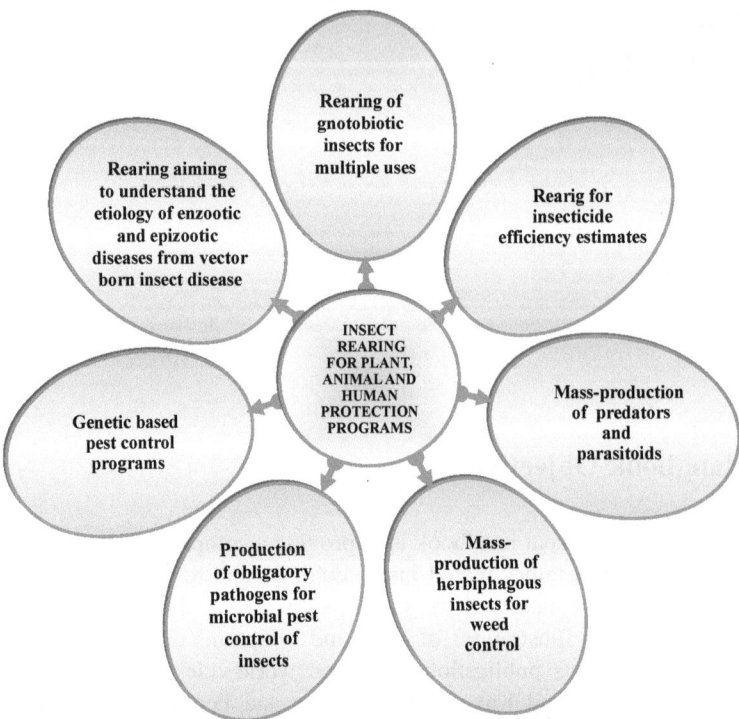

Fig. 1.1 Primary objectives for insect rearing

1.2 Reared Insects in Biotechnology Programs

The mass-production of different species of insects shows at present time a growing ecological interest and a relevant economic impact. Apart from the traditional rearing in apiculture and silkworm production systems, a pertinent necessity to rear a varied numbers of insect species in laboratory conditions, including both phytophagous and entomophagous species, also exists (Van Driesche et al. 1996). In particular, insect mass-production technologies aiming for the protection of valuable crops in artificial or natural environments (e.g., greenhouses, agriculture and forestry ecosystems), as well as the protection of human populations from insects vector of disease, are well established (Fig. 1.1). The most prevalent direction in pest control activities has been the biological control of noxious species of plants and invertebrates based in the mass-production of beneficial insects. Invasive insect species have attained especial relevance due to their effects in the ecological balance of indigenous biological communities in both agriculture and forestry ecosystems, as well as impacts in human health.

Classical biological control aiming the introduction and establishing of phytophagous, predatory or parasitic invertebrates, from the native origin of the invasive pests, is the primary tool in the current management strategies aiming to halt the devastating effects of alien noxious species (Van Driesche et al. 2010). These pest control strategies are directly connected with the mass-rearing of useful exotic invertebrates, namely insects. Strong health control measures, for any potential introductory species, should always be enforced since these species can be vectors of pathogenic microorganisms which could potential strive in new environments, which unpredictable consequences. Thus, it is imperative to implement diagnostic and prophylactic protocols to detect infectious diseases in arthropod populations reared in laboratory or industrial facilities, prior to release, hence eliminating collateral detrimental effects from the introduction of exotic arthropods in biocontrol programs (Lenteren et al. 2006, Lenteren and Loomans 2006).

Gnotobiotic insects, i.e., "sterile" insects born in aseptic laboratory conditions, and commonly used for the study of etiology of disease, and for the cultivation of axenic cells, tissues and organs in vitro, are instrumental in the development of biocontrol programs ensuring the inexistence of cryptic pathogens in the insect colonies to be release in a given environment. This is especially important when studying poikilothermic invertebrates (i.e., organisms with body temperature oscillations according to ambient temperature) since they typically present unstable microbiota which can significantly influence the animal's behavior according to different environmental conditions (Talpalatsky et al. 1984; Gouli et al. 1999).

In microbiological pest control programs gnotobionts can also be used for the serial passage of microbial pathogens through the host in order to maintain, or increase virulence. For example, if entomopathogenic fungi are continuously cultivated on artificial media they tend to lose their original levels of pathogenicity. However, if these isolates are routinely inoculated into the original host, pathogenicity is often restored (Tanada and Kaya 1992). In addition, living long-term storage of pathogens can effectively be done in gnotobiotic insects. For example, the gnotobiotic greater wax moth, *Galleria mellonella*, is used for long-term storage of industrial strains of the entomopathogenic bacterium *Bacillus thuringiensis* (Ivanov et al. 1984). The mass production of viruses, rickettsiae and protozoans in gnotobiotic animals is also viable using living biota as storage hosts. We can also forecast that gnotobiotic organisms will probably be instrumental in the search for new colonization frontiers, namely space, as it is imperative to avoid undesirable biological contaminants in unknown habitat conditions.

1.3 Importance of Insect Infectious Diseases

Infectious diseases are broadly distributed in all biological kingdoms and are a strong regulatory factor in natural animal populations. There are many historical examples showing the critical importance of infection diseases in insect rearing. The silkworm breeding industry in France was completely destroyed in the nineteenth century as a result of microsporidia, *Nosema bombycis*, infections. The famous French scientist Luis Pasteur solved this problem developing special analysis techniques to assess insect pathologies, selecting only healthy individuals for rearing colonies. More recently, infections with microsporidia have also been reported (Van Frankenhuyzen et al. 2004). Disease screening techniques are well established in apiculture, where quality honey bee products are only produced if a complex of disease control and prophylactic measures are taken, in order to warrant the health of the populations. Successful mass-production of any arthropod species can only be achieved with pre existent basic knowledge of causal agents of disease for that particular insect species and the recognition of its signs and symptoms in the host organism(s).

1.4 Economic Impact of Insects

The risk for accidental introduction of invasive species has increased, in parallel with the rise of global trade. Sea and air transportation are the two principal paths for undesirable species invasion. The volume of transported goods is increasing dramatically. For the period of 1990–2000, the number of cargo containers arriving to the USA doubled. Unfortunately, the number of non native insect species that arrive to the continent through this pathway also raised (Work et al. 2005).

In the last years, the numerous invasive noxious species introduced in the American continent have created important ecological and economical problems (Pimentel et al. 2000, 2001, 2005). The estimated number of non-indigenous biological species in the US is approximately 50,000. They account for annual losses of approximately $137 billions a year (Pimentel et al. 2000). The *Diabrotica* spp. insect complex for instance, accounts for $1 billion of annually costs in the US (Al-Deeb and Whilde 2005). The production costs attributed to scale insects (Homoptera: Coccoidae) in the US reach $500 million per year (Kosztarab 1996) and the ecological and economical cost of the exotic weed purple loosestrife, *Lythrum salicaria*, reach $45 millions per year (Blossey et al. 2001). These are not marginal values.

1.5 Classification of Insect Diseases

Insect diseases can be divided into two principal groups: non-infectious and infectious diseases (Fig. 1.2). Both disease complexes have significant relevance in insect rearing. Non-infectious diseases can easily be managed since, as a rule, they are directly correlated with the physiology of the insect species and behave opportunistically when physiological pressures affect suitable hosts, which then become more prone to be infected with pathogens, or even saprophytic microorganisms. For example, numerous bacteria living in the insect alimentary canal, as natural components of the microbiota, can be active pathogens if the hosts become physiological stress. Extreme physical or chemical situations impose to an

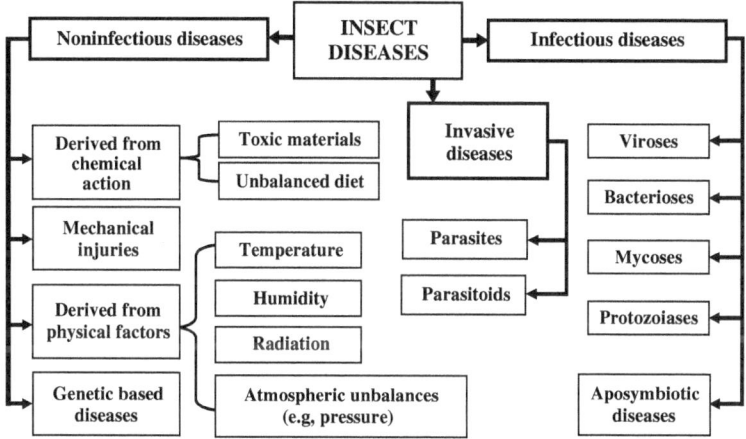

Fig. 1.2 Classification of insect diseases

organism can activate latent virus infections in suitable hosts. Thus, the maintenance of optimal physical breeding conditions, as well as the assurance of all
nutrition requirements for mass-rearing of insects, are essential factors for successful laboratorial and industrial mass rearing of insects.

References

Al-Deeb MA, Whilde GE (2005) Effect of Bt corn expressing the Cry3Bb1 toxin on western corn
rootworm (Coleoptera: Chrysomelidae) biology. J Kans Entomol Soc 78:142–152

Blossey B, Skinner L, Taylor J (2001) Impact and management of purple loosestrife (*Lythrum
salicaria*) in North America. Biodivers Conserv 10:1787–1807

Boucias DG, Pendland JC (1998) Principles of insect pathology. Kluwer Academic Publishers,
Boston/Dordrecht/London

Gouli VV, Ribina SY (1988) Viral infections of insects and their diagnostics. Stiinza C (ed)
(Russian)

Gouli VV, Parker BL, Reid W (1999) Method for rearing gnotobiotic thrips. J Appl Ent
123:127–128

Hajek AE (2004) Natural enemies: An introduction to biological control. Cambridge University
Press, UK

Ivanov GM, Gouli VV, Mikhalev AL (1984) The use of gnotobiotic insects for the maintenance
of bacterial cultures of *Bacillus thuringiensis* Berl. Group. Collected articles of the All-Union
Institute of Biological Plant Protection: Microorganisms in Plant Protection, Kishinev,
Moldova: 31-33 (In Russian, summary in English)

Kosztarab M (1996) Scale insects of northeastern North America. Virginia museum of natural
history special publication number 3

Lacey LA (1997) Manual of techniques in insect pathology (Biological techniques series).
Academic Press, NY

Lacey LA, Kaya HK (2007) Field manual of techniques in invertebrate pathology: application
and evaluation of pathogens for control of insects and other invertebrate pests. Springer, NY

Lenteren JC, Loomans AJM (2006) Environmental risk assessment: methods for comprehensive evaluation and quick scan. In: Bigler F, Babendreier D, Kuhlmann U (eds) Environmental impact of invertebrates in biological control of arthropods: methods and risk assessment. Wallingford, UK

Lenteren JC, Bale J, Bigler F, Hokkanen HMT, Loomans AJM (2006) Assessing risks of releasing exotic biological control agents of arthropod pests. Annu Rev Entomol 51:609–634

Pimentel D, Lach L, Zuniga R, Morrison D (2000) Environmental and economic costs of nonindigenous species in the United States. Biosci 50:53–65

Pimentel D, McNair S, Janecka J, Wightman J, Simmonds C, O'Connell C, Wong E, Russel L, Zern J, Aquino T, Tsomondo T (2001) Economic and environmental threats of alien plant, agriculture, animal and microbe invasions. Agric Ecosys Environ 84:1–20

Pimentel D, Zuniga R, Morrison D (2005) Update on the environmental and economic costs associated with alien-invasive species in the United States. Ecolog Econ 52:273–288

(Jr.) Poinar GO, Tomas GM (1984) Laboratory guide to insect pathogens and parasites. Springer, NY

Solter LF, Becnel JJ (2000) Entomopathogenic microsporidia. In: Lacey L, Kaya H (eds) Field manual of techniques for the evaluation of entomopathogens. Kluwer Academic Publishers, Dordrecht, The Netherlands

Talpalatsky PL, Mikhaelev AI, Gouli VV, Ivanov GM, Sukharukova AN (1984) Gnotobiotic insects as a model to study experimental pathology. Collected articles of the All-Union Ins. Biol. Pl. Protection: Mass-rearing of insects, All-Union Institute Kishinev, Moldova 5-16 (In Russian, summary in English)

Tanada Y, Kaya H (1992) Insect pathology. Academic Press, NY

Van Driesche RG, Healy S, Reardon R (1996) Biological control of arthropod pests of the northeastern and north central forests in the United States: a review and recommendations. FHTET-96-19, US Department of Agriculture, Forest Service, Morgantown, West Virginia

Van Driesche et al (2010) Classical biological control for the protection of natural ecosystems. Biol Control 54:S2–S33

Van Frankenhuyzen K, Ebling P, McCron B, Ladd T, Gauthier D, Vossbrink C (2004) Occurrence of *Cystosporogenes* sp. (Protozoa, Microsporidia) in a multi-species insect population facility ans its elimination from a colony of eastern spruce budworm, Choristoneura fumiferana (Clem.) (Lepidoptera: Tortricidae). J Inv Pathol 87:16–28

Work T, McCullough D, Cavey J, Komsa R (2005) Arrival rate of nonindigenous insect species into the United States through foreign trade. Biol Invasions 7:323–332

Chapter 2
Principal Group of Infectious Insect Diseases

Abstract Provides a comprehensible and copious illustrated description of the most common diseases in laboratory reared insect colonies, comprising Viruses (Baculoviridae, Reoviridae, Poxviridae, and Iridoviridae); Bacteria (Bacillaciae, Pseudomonadaceae and Enterobacteriaceae); Rickettsia (Rickettsiaceae); Protozoa (Amebiases, Gregarine and Coccidian); Fungi (Entomophthorales and Muscardine) and Microsporidia. The morphology of the pathogen, its external signs and symptoms in infected hosts, the pathomorphism of disease and the most common insect host for each pathogen group are described. Descriptions for the different species or families of microorganism agents of disease varied according to the current scientific knowledge of the organisms being described, its pathogenic importance and management potential.

Keywords Insect colonies · Pathogens · Signs · Symptoms · Virus · Bacteria · Rickettsia · Protozoa · Fungi · Microsporidia

All insect infectious diseases are classified according to the nature of the pathogenic organism. The principal groups of infectious diseases comprise viruses, bacteria, fungi and protozoa. Insects can manifest disease caused by one type of microorganisms or a complex of microorganisms. Infections based on two different types of viruses, viruses and fungi or fungi and bacteria are common. However, as a rule, a specific pathogen triggers the infection process. The immune-compromised organism is then suitable for the influence of semi-saprophytic or saprophytic microbes, which take advantage of the compromised defenses of the host. Specific pathogens exist in all groups of microorganisms, but entomopathogenic viruses and protozoa can be distinguished by their high level of parasitism. Often, these pathogens can cause limply-flowing infection processes in laboratory reared insect populations, subjecting the insects to a high physiological stress, which ultimately opens the path for facultative entomopathogenic microorganisms, such as bacteria or

V. Gouli et al., *Common Infectious Diseases of Insects in Culture*,
SpringerBriefs in Animal Sciences, DOI: 10.1007/978-94-007-1890-6_2,
© Vladimir Gouli 2011

fungi, to invade the insect body. Insects are also associated with symbiotic microorganisms, depending on them for critical physiological processes. This fact is extremely important as artificial diets for laboratory reared insects contain, in some cases, antibiotic elements which can suppress the activity of the symbionts, hence affecting the insect's normal develop and ultimately causing death.

2.1 Viral Diseases

2.1.1 Typical Nucleopolyhedroses

There are numerous viruses associated with different groups of insects. Viruses are distributed in 15 families. The main insect pathogens are found in the families of Baculoviridae, Reoviridae, Poxviridae and Iridoviridae. Among these families the Baculoviridae are responsible for the most important insect infections. Commonly, the Baculoviridae are subdivided in three genera: (a) nucleopolyhedrosis viruses, (b) granulosis viruses and (c) nonoccluded, rod-shaped nuclear viruses.

The nucleopolyhedrosis and granulosis viruses are widely distributed in natural insect populations. These pathogenic microorganisms are very important factors regulating insect population effectives. Laboratory reared Lepidoptera and some Hymenoptera species are very susceptible to these two groups of viruses. Viruses can maintain their infective potential for several years, without population outbreaks, and in the most diverse environmental conditions. These biological traits facilitate the horizontal transmission of pathogens in insect populations. At present time, several hundred insects are registered as hosts of baculoviruses (e.g. Reardon et al. 1996). The nonoccluded, rod-shaped nuclear viruses are known to produce infection in Coleoptera, Homoptera, Diptera, Lepidoptera, and Hymenoptera insects (Tanada and Kaya 1992). The viruses responsible for cytoplasm polyhedroses (family of Reoviridae), pox diseases (family of Poxviridae) and iridescent viral diseases (family of Iridoviridae) have seldom occurrence patterns. The most important viral diseases in laboratory insect mass-reared insects have very specific symptoms and simple diagnostic methods can be established to recognize each type of disease. General information about viral diseases with significant importance during insect mass-rearing are described and depicted next.

2.1.1.1 Morphology of the Pathogens

Typical nucleopolyhedrosis viral diseases consist of DNA containing rod-shaped viruses particles which infect and replicate in infected cells of the host. Size of virus particles range from 40–60 nm to 200–400 nm. These submicroscopic structures are impossible to observe with the naked eye but fortunately, as a rule, the virus particles are occluded in relatively big crystal polyhedral structures (polyhedra) with sizes ranging from 0.5 μm to 15 μm (Fig. 2.1) which significantly

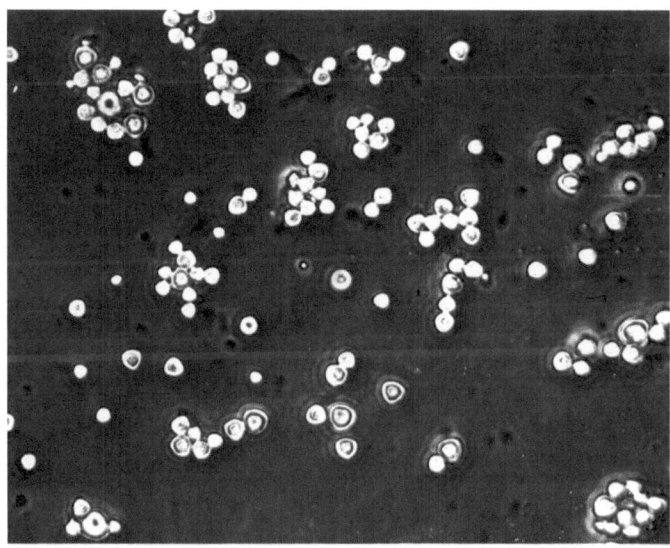

Fig. 2.1 Baculovirus polyhedral inclusions (capsules in infected cells) under simple light microscope. Phase contrast, objective 40x

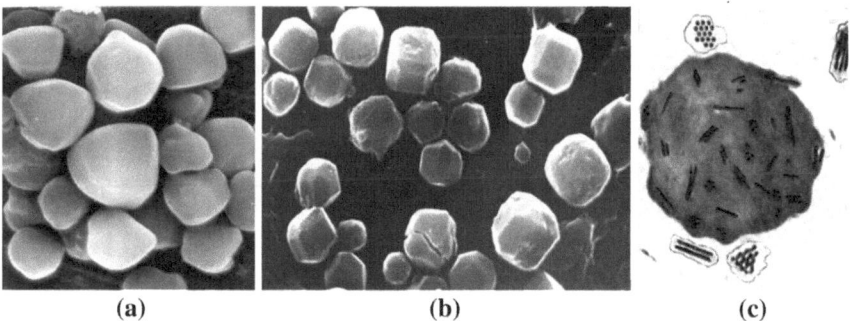

       (a)   **(b)**   **(c)**

Fig. 2.2 Baculovirus polyhedral inclusions from **a** Tent caterpillar moth, *Cosmotriche lunigera,* and **b** Cabbage moth, *Mamestra brassicae.* Scanning electron microscopy, 15,000x; **c** Thin section of polyhedron from nun moth, *Lymantria monacha,* 60,000x

facilitate the disease diagnostic since they are visible under a standard light microscope. Scanning electron microscope showing morphology of different types of polyhedra can provide informative characters enabling viral species identification but they are not indispensable for disease identification (Fig. 2.2a–c).

2.1.1.2 External Signs and Symptoms of Disease

Insects lack appetite and decrease activity. This type of disease develops very quickly. The most characteristic external sign of nuclear polyhedroses is the white or yellow color of the insect integument. The body is completely filled with viral

(a)   (b)   (c)   (d)

Fig. 2.3 Typical position of insect cadavers after nuclear polyhedrosis infection. **a** Tropical unidentified Lepidoptera larvae; **b** Broom moth, *Mamestra pisi*; **c** Cabbage moth, *Mamestra brassicae*; **d** Gypsy moth, *Lymantria dispar*

Fig. 2.4 Greater wax moth, *Galleria mellonella* with signs of nuclear polyhedroses

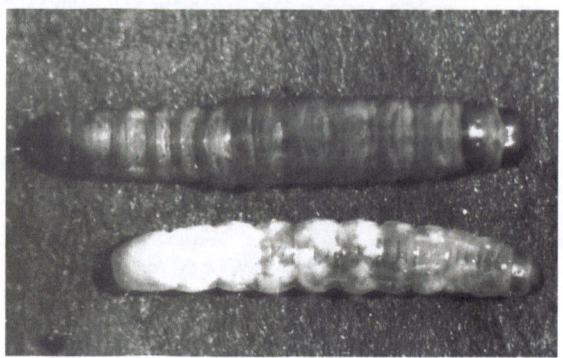

inclusions, and the dead insects can be easily destroyed at physical contact since the hypoderm is easily shattered. Corps usually show specific positions and can appear perched on the upper parts of insectariums (Fig. 2.3a–d). Insects having non-pigmented or semi-transparent cuticle present a fat body with extensive white zones corresponding to the accumulation of nuclear polyhedra (Fig. 2.4). Development of disease is slowed down at metamorphosis. Larvae infected in last instars can still develop to adult insects. As a rule the resulting adults present morphological anomalies (Fig. 2.5).

2.1.1.3 Pathomorphology of Disease

Usually, viruses attack the fat body, hypoderm, tracheal matrix, hemocytes and others cells (Fig. 2.6a–c). The formation of polyhedra takes place in cell nuclei. As a result, the nuclei show high hypertrophy and practically occupies all the cell space. In an advanced stage of the disease, the nuclei are destroyed and polyhedral inclusions are released outside the cell. During this period of disease it is possible

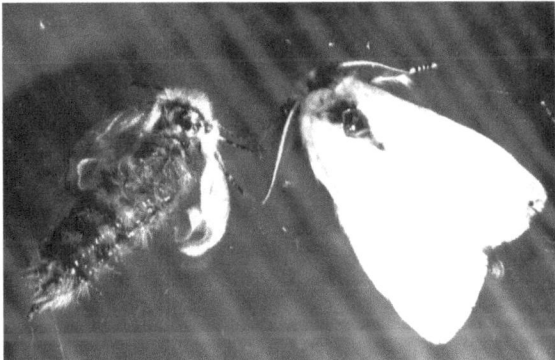

Fig. 2.5 Fall webworm, *Hyphantria cunea* with morphological wing anomalies caused by nuclear polyhedroses viral infection

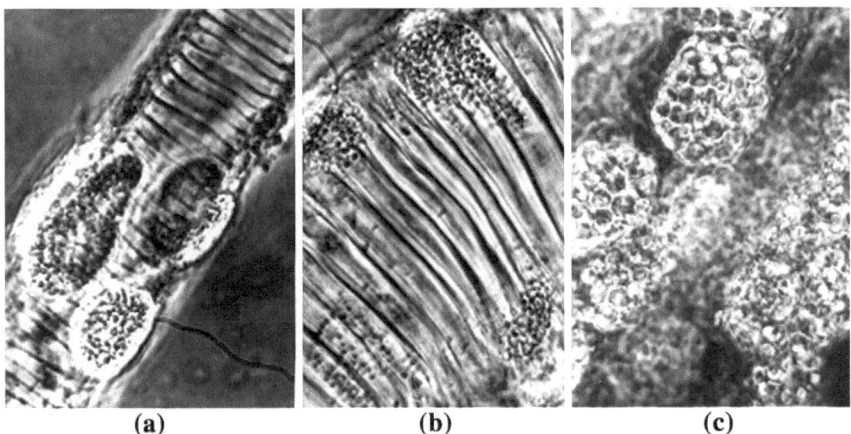

(a) (b) (c)

Fig. 2.6 Nuclear polyhedrosis infection of Peppered moth, *Biston betularia*. **a** Hypertrophy of tracheal epithelium cells, phase contrast, objective 40x; **b** Cell lyses, phase contrast, objective 40x; **c** Fat body, phase contrast, objective 90x

to locate viral inclusions in the cytoplasm of hemocytes as a result of phagocytosis processes (Fig. 2.7a, b). In the final period of the disease the insects' hemolymph presents a specific milky color as a result of cell destruction and the dissemination of polyhedral to the hemolymph.

2.1.1.4 Hosts

This type of diseases is very common in Lepidopteran species. Susceptible hosts are also found in Diptera, Coleoptera and some others insect orders. However, more than 95% of registered arthropod diseases around the world are related to

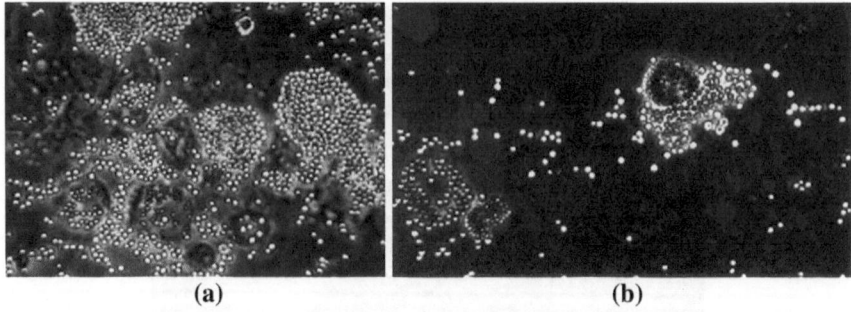

(a)   (b)

Fig. 2.7 Insect hemocytes in the last stage of nuclear polyhedrosis presenting numerous viral inclusion bodies. **a** Hemocytes of gypsy moth, *Lymantria dispar*, phase contrast, objective 40x; **b** Peppered moth, *Biston betularia*, dark field, objective 40x

Fig. 2.8 Baculovirus polyhedral inclusions from European pine sawfly, *Neodiprion sertifer*. Scanning electron microscopy, 15,000x

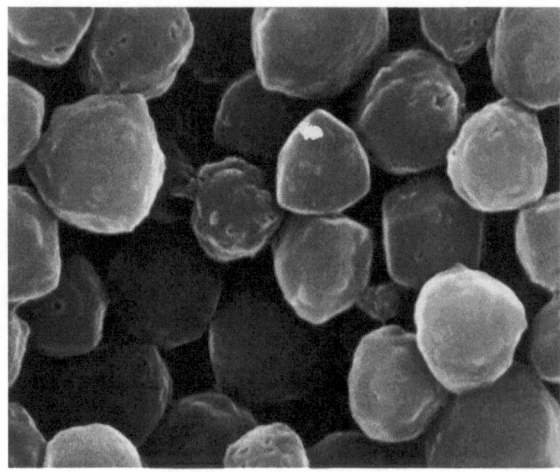

insect species in the Lepidoptera, especially in the families of Lymantriidae, Lasiocampidae, Tortricidae and Geometridae.

2.1.2 Nuclear Polyhedroses of the Intestinal Epithelium

2.1.2.1 Morphology of the Pathogens

Viruses causing this type of disease have similar morphology to the viruses causing typical nucleopolyhedroses. However, the size of the virus particles is slightly smaller with approx. 50 nm × 250 nm. Single viral particles are located in crystal polyhedral structures. The polyhedra size ranges from 0.5 μm to 5 μm (Figs. 2.8, 2.9).

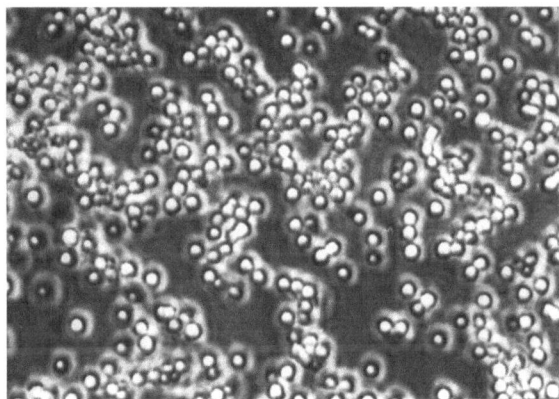

Fig. 2.9 Baculovirus polyhedral inclusions from Poplar sawfly, *Trichiocampus viminalis* under light microscope. Phase contrast, objective 40x

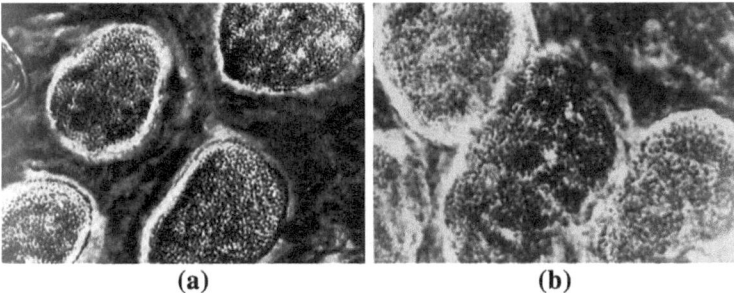

(a)   (b)

Fig. 2.10 **a** Early and **b** Late stages of hypertrophy in nuclei of intestinal epithelium cells from the European pine sawfly, *Neodiprion sertifer*. Phase contrast, objective 40x and 70x, respectively

2.1.2.2 External Sign and Symptom of Disease

The insects show lack of appetite. Colony species loose group instincts. Infected larvae may present a white color. Insects excrete brown liquid flows from the anus, which can dry and fix the insect to the substratum. Often, diseased larvae show a white liquid flowing from the mouth which contains numerous viral inclusions. The diseases develop rapidly (3–5 days) and death overcomes in a short period of time.

2.1.2.3 Pathomorphology of Disease

The viruses attack only the intestinal epithelium. The formation of polyhedra takes place in cell nuclei. Nuclei show a high hypertrophy and practically occupy all the cell volume (Fig. 2.10a, b). The nuclei are quickly destroyed and polyhedral inclusions find their way into the gut and sometimes into the hemolymph. During this period of disease it is possible to find the virus inclusions in hemocytes, as a result of phagocytosis, and in excrements. As a rule, the development of the

(a)   (b)

Fig. 2.11 Adult female of European pine sawfly, *Neodiprion sertifer* with **a** Encapsulated and melanized group of cells; **b** Thin section of abdomen with melanized group of cells

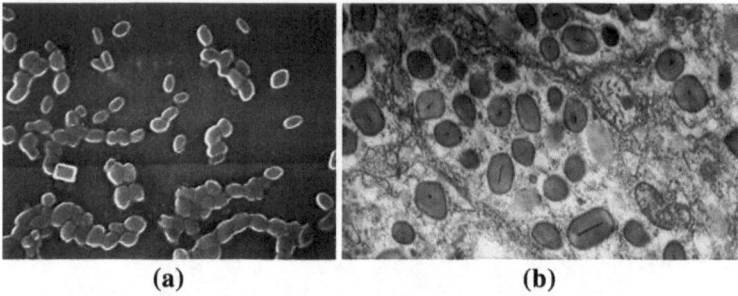

(a)   (b)

Fig. 2.12 **a** Granules from *Agrotis segetum*. Scanning electron microscope, 10 000x; **b** Thin section of cell infected with granulosis virus from *A. segetum*. Scanning electron microscopy, 25,000x

disease is halt during insect metamorphosis. During this phase the larval midgut acquires the adult form and groups of encapsulated and melanized infected cells reallocate in the body cavities. Sometimes these formations can be visible through the cuticle of adult insects (Fig. 2.11a, b).

2.1.2.4 Hosts

Nuclear polyhedroses of the intestinal epithelium are known to infect mainly sawfly species (Hymenoptera: Diprionidae, Tenthredinidae).

2.1.3 Granuloses

2.1.3.1 Morphology of the Pathogens

Viruses form ellipsoidal inclusions, i.e., capsules in the infected cells. The capsules are much smaller than the different types of polyhedra and vary in size from 120 nm to 300 nm and from 300 nm to 500 nm (Fig. 2.12a). The virus inclusions can be recognized under simple light microscope only after special methods of staining. As a rule, the capsule contains a single rod-shaped virus particle (Fig. 2.12b).

2.1.3.2 External Signs and Symptoms of Disease

The diseased insects have the same symptoms as typical nucleopolyhedroses.

2.1.3.3 Pathomorphology of Disease

The viruses attack the fat body, hypodermal cells, and epithelium of tracheal matrix. The formation of granule takes place in both cell nuclei and cytoplasm. The nuclei are hypertrophied and can be completely destroyed. The granules can be recognized with the help of staining methods. These are described in Chap. 3.

2.1.3.4 Hosts

Granulosis are widely distributed in *Lepidoptera* species, especially in the Noctuidae family.

2.1.4 Cytoplasmic Polyhedroses

2.1.4.1 Morphology of the Pathogens

Cytoplasmic polyhedroses are acute diseases caused by viruses related to the Reoviridae family. RNA contained viruses form polyhedral inclusion bodies, polyhedra, in the cytoplasm of epithelial cells in the hosts' midgut. The size of polyhedra is very broad ranging from 0 to 5 µm to 10 + µm (Fig. 2.13). The polyhedra contain numerous icosahedral-shaped virus particles with a size from 50 nm to 80 nm.

2.1.4.2 External Signs and Symptoms of Disease

The diseased insects show symptoms closely related to nuclear polyhedrosis of the intestinal epithelium. Since viruses develop in the midgut, the polyhedra are released into the gut lumen and excreted with feces (Fig. 2.14). The caterpillar is fixed to the substratum through intestinal excretions containing practically only pure polyhedra. Often such insects survive some time and try to move. As a result, we can see signs of excretion scattered on the substratum surface.

Fig. 2.13 Cytoplasmic
polyhedra from gypsy moth,
Lymantria dispar. Scanning
electron microscopy, 15,000x

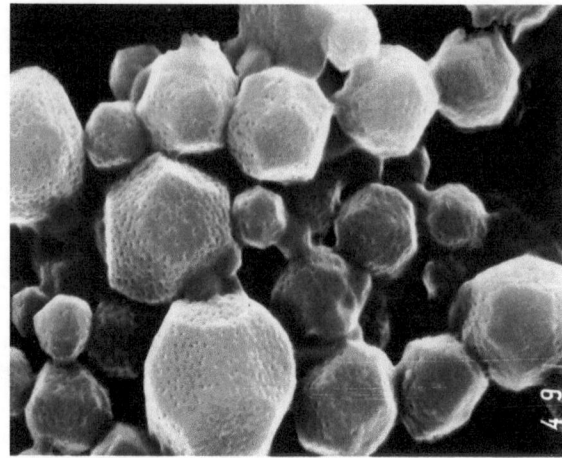

Fig. 2.14 Intestinal
excretion of garden tiger
moth caterpillar, *Arctia caja*,
infected with cytoplasmic
polyhedral virus

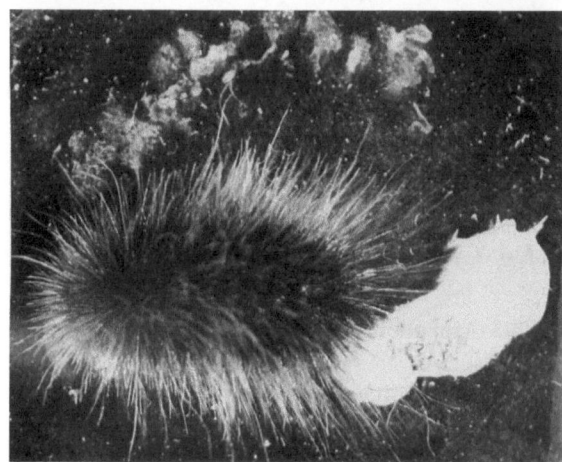

2.1.4.3 Pathomorphology of Disease

The virus attacks only the midgut epithelial cells. Polyhedra formation takes place
in the cytoplasm of cells. The cells show hypertrophy and are completely
destroyed. The midgut of diseased larvae becomes thick and white (Fig. 2.15).

2.1.4.4 Hosts

Cytoplasmic polyhedroses are distributed mainly within the order Lepidoptera, and
to a less extend in Diptera, Coleoptera and Neuroptera. The first report of a disease
caused by a reovirus was in silk worm hosts, *Bombyx mori,* in Japan (Aruga
and Tanada 1971). Cytoplasmic polyhedroses are seldom manifested in natural
population of insects. Our long time survey for entomopathogenic viruses in

Fig. 2.15 Midgut of garden
tiger moth, *Arctia caja*
infected with cytoplasmic
polyhedral virus

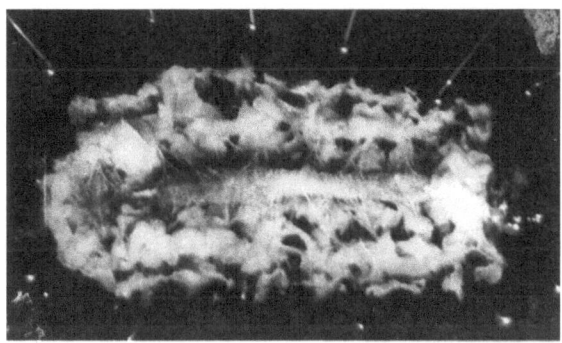

European–Asian insect populations enabled us to record approx. 100 insect species susceptible to nuclear polyhedroses and only several species vulnerable to cytoplasmic polyhedroses. However, this group of infections can play an important role in insect mass reared populations. We have reared several insects including gypsy moth, *Lymantria dispar*, cabbage moth, *Mamestra brassicae*, and cotton moth, *Helicoverpa armigera*, due to spontaneous manifestation of cytoplasmic polyhedroses in laboratory reared populations, in order to retrieve baculoviruses suitable to be used as alternative biological insecticides.

2.1.5 Poxvirus Diseases

2.1.5.1 Morphology of the Pathogens

Poxvirus diseases are infections caused by viruses in the Poxviridae family. Poxvirus form large protein bodies with a size from 1 μm to 15 μm, showing ovoid, spindled and rhomboid forms (Fig. 2.16a). The protein bodies contain viral particles with a brick shape or oval forms and occupy the entire inclusions (Fig. 2.16b). Poxvirus pathogens are generally found in cockchafer, *Melolontha melolontha*.

2.1.5.2 External Signs and Symptoms of Disease

Insects infected with poxviruses present a white coloration. The larvae grow to more than the double of their normal size and weight. Diseased insects are usually active, and often die in the oldest instars or during metamorphosis.

2.1.5.3 Pathomorphology of Disease

Viral replication takes place mainly in the fat body (Fig. 2.17), although the pathogen can also be present in others organs and tissues. Inclusion bodies are formed in the cell's cytoplasm. The cells are hypertrophed and ultimately destroyed.

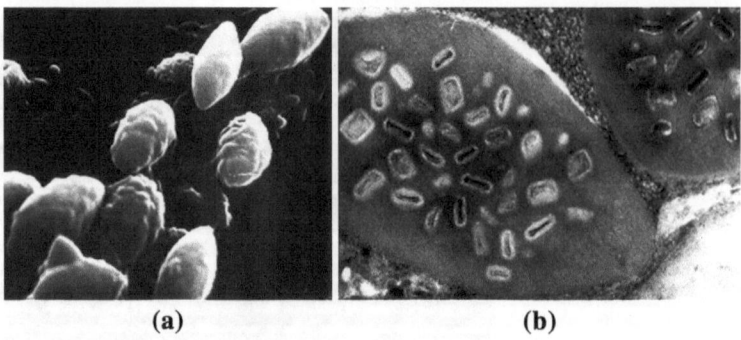

(a)   **(b)**

Fig. 2.16 a Poxvirus inclusion bodies from the buzzer midge, *Chironomus plumosus*, Scanning electron microscopy, 10,000x; **b** Section of cell infected with poxvirus from *Ch. Plumosus*. Scanning electron microscopy, 50,000x

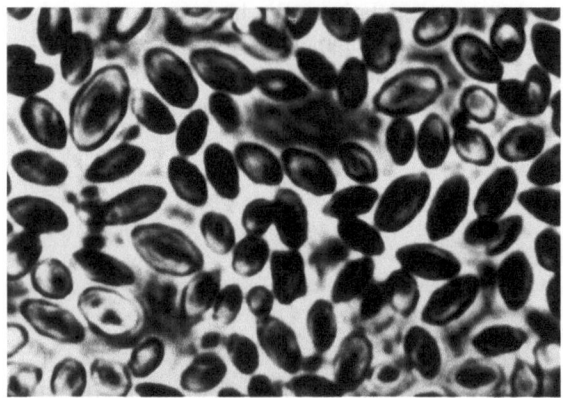

Fig. 2.17 Poxvirus inclusions in fat body of the buzzer midge, *Chironomus plumosus*. Light microscope, phase contrast, objective 90x

2.1.5.4 Hosts

The family of Poxviridae includes numerous viruses responsible for diseases of many invertebrates and vertebrates, including man. Entomopathogenic poxviruses are found in the orders Coleoptera, Lepidoptera, Diptera, Orthoptera, and Hymenoptera.

2.1.6 Iridescent Viral Diseases

2.1.6.1 Morphology of the Pathogens

Iridescent virus diseases are infections caused by viruses in the Iridoviridae family. There are two iridescent virus genera causing disease in insects, *Iridovirus* and *Chloriridovirus*. The iridoviruses present an icosahedral form or virions. Viruses

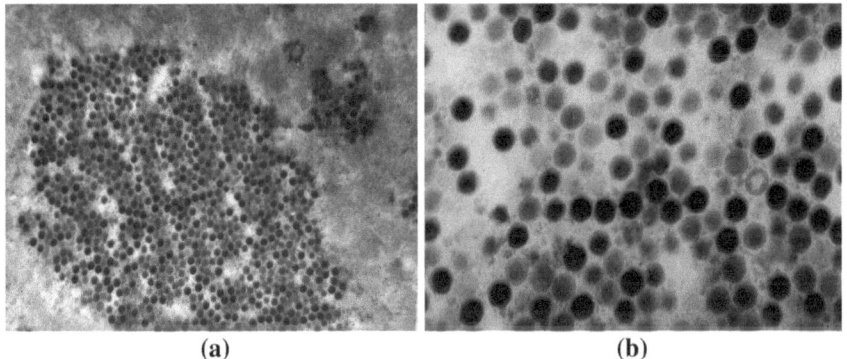

(a) (b)

Fig. 2.18 Iridovirus from *Chironomus plumosus*. A Scanning electron microscopy, 10,000x (**a**), and 80,000x (**b**), respectively

from the genus *Iridovirus* present a size ranging from 120 nm to 130 nm in diameter, and the genus *Chloriridovirus* approximately 180 nm in diameter (Fig. 2.18). The viral particles form microcrystals which are responsible for specific iridescence patterns.

2.1.6.2 External Signs and Symptoms of Disease

The initial signs of infection are linked with the appearance of iridescence. Iridescence is easily visible in the case of nonpigmented insects. In insects with pigmented integument the iridescence can be observed in the inter-segmental membranes. The development of disease is associated with the appearance of white spots on the insects' body. In infected insects the iridescence presents a yellow–green or violet–blue color under ultraviolet light.

2.1.6.3 Pathomorphology of Disease

Iridoviruses cause systemic infections involving different organs and tissues, however the insect fat body is the principal place for viral development. Often, the viral particles form regular crystallized structures which are visible under a light microscope.

2.1.6.4 Hosts

Iridoviruses occur in the orders Diptera, Coleoptera, Lepidoptera, Hymenoptera, Hemiptera and Orthoptera. The majority of this type of diseases has been described in Diptera species with aquatic and unpigmented larvae.

2.2 Bacterial Diseases

There are several hundred species of bacteria associated with insects. Most of them are common microbiota of the digestive tract of insects, some are facultative entomopathogens, and a few are obligate pathogens. The pathogenic bacteria usually infect an insect through the digestive tract using specific toxins to enable penetration through the gut into the body cavity. The most important entomopathogenic bacteria are located in the families of Bacillaciae and Enterobacteriaceae, however, facultative pathogens can be found in the families of Pseudomonadaceae, Vibrionaceae, Micrococcaceae, and Streptococcaceae, among others. Typically, bacterial infections agents exist in the family of Bacillaciae, genera *Bacillus* and *Clostridium*. Usually, the most common diseases are caused by the species: *Bacillus thuringiensis, B. cereus, B. popilliae, Clostridium brevifaciens,* and *C. malacosomae.*

2.2.1 *Bacterioses Caused by* **Bacillus thuringiensis** *and* **Bacillus cereus**

2.2.1.1 Morphology of the Pathogens

Both species of bacteria are grampositive and present similar morphology. Dimensions vary from 0.9 μm to 1.4 μm wide and from 3.0 μm to 7.0 μm long. Bacterial cells are movable and often form chains. Bacterial spores are formed centrally or sub centrally in the cell (Fig. 2.19). Spores without sporangium have oval form. *B. thuringiensis* comprises more than 30 different subspecies, differentiated by their biochemical properties, endotoxin morphology and host-specific pathogenicitiy. *B. thuringiensis* produces a crystalloid toxin (also called crystal inclusion, crystal endotoxin or parasporal body) which allows the morphological differentiation from *B. cereus* (Fig. 2.20). Although the endotoxin can present different morphologies, typically there are two types of crystals, bipyramidal and ovoidal.

2.2.1.2 External Signs and Symptoms of Disease

Bacterioses develop during a short period of time. Insects stop feeding, loose mobility, rapidly darken and miss turgor. The cadavers usually present a saprogenic odor. Dead insects shrivel and harden. The cadavers become focal sources for living spores.

2.2.1.3 Pathomorphology of Disease

The most important entomopathogen is the bacterium *B. thuringiensis.* Infection process progresses through the digestive tract. Bacterial endotoxin spread under the influence of intestinal enzymes destroying the insects' gut wall. The bacteria

Fig. 2.19 Group of sporulated cells of *B. cereus.* Scanning electron microscopy, 10,000x

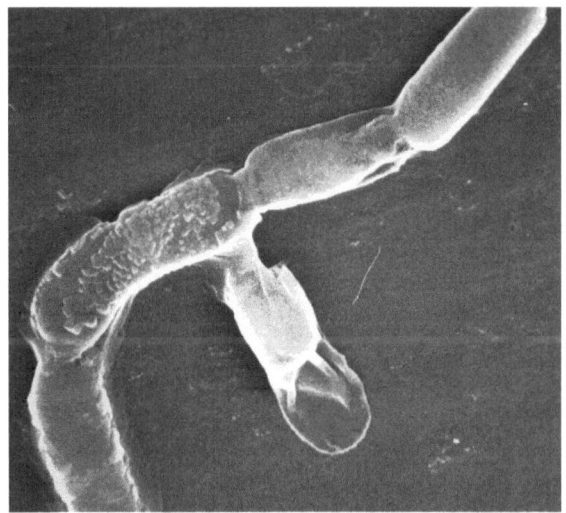

Fig. 2.20 *B. thuringiensis* spores and endotoxin crystals. Scanning electron microscopy, 10,000x

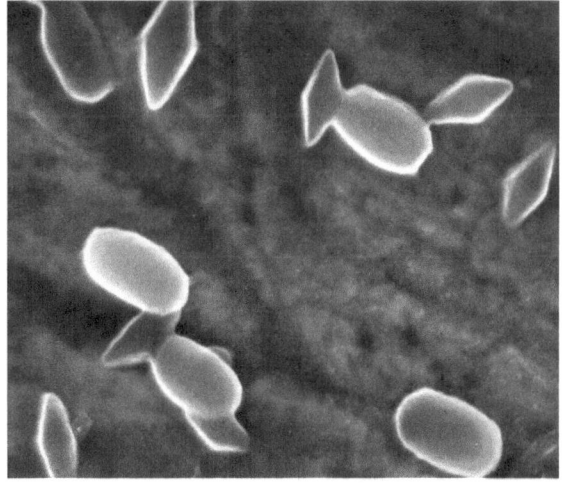

proliferate through the body cavity and very quickly colonize all the organs and tissues of the host. The insect's body becomes a knap-sack with bacteria and finally turns black (Fig. 2.21). Infection process can be trigger as a result of mechanical or others injuries in the cuticle. The initial lesion becomes dark in consequence of bacterial invasion (Fig. 2.22).

2.2.1.4 Hosts

B. cereus and *B. thuringiensis* have a worldwide distribution. Insects from the orders Lepidoptera, Diptera, and Coleoptera show high susceptibility levels to host specific subspecies of *B. thuringiensis*.

Fig. 2.21 Healthy and dead
(*black*) *Pieris brassicae*
pupae after bacterial infection

Fig. 2.22 Coleopteran larva
with local percutaneous
bacterial infection

2.2.2 Milky Diseases Caused by the Bacteria Bacillus popilliae and Bacillus lentimorbus

2.2.2.1 Morphology of the Pathogens

B. popilliae causes milky disease of scarabaeid insects, known as type A disease,
and *B. lentimorbus* causes type B disease. The bacterium *B. popilliae* is movable,
and can be stained with Gram's stain. The size of bacterial cells range from 0.9 μm
to 1.0 μm wide and 5.0–8.0 μm long. Spores form in the middle section of cells.
Simultaneously with spore formation, bacterial cell form parasporal inclusions
associated with the spore, which has a truncated cone form. The size of the
parasporal inclusion ranges from 0.7–0.8 μm × 0.5–0.6 μm. After sporulation
the cells present hypertrophy since spores are larger than the sporangium.
The bacterium *B. lentimorbus* is morphologically closely related to *B. popilliae*,
however, the size of the bacterial cells is smaller, ranging from 0.5–0.7 μm to
1.8–7.0 μm. The bacterium does not produce parasporal inclusions.

2.2.2.2 External Signs and Symptoms of Disease

Infected insects present a sluggish consistency, the larvae usually do not molt and metamorphosis does not occur. The infected larvae turns white due to the high number of bacteria contained in the insect's hemolymph. In case of *B. popilliae* infections (type A disease), the larvae maintains a white coloration for a long period of time, whereas larvae infected with *B. lentimorbus* becomes very dark after a short period of time.

2.2.2.3 Pathomorphology of Disease

The bacterial spores are carried to the insect's digestive tract together with food. The spores germinate, penetrate into midgut cells, and invade hemocoel cavities. Eventually the bacterial cells fill the insect's body completely. The development of the disease progresses during a long period of time since these bacteria do not produce toxic substances.

2.2.2.4 Hosts

Milky disease has been identified in scarabaeids beetles in different geographical zones around the world.

2.2.3 Pseudomonadaceae and Enterobacteriaceae Infections

Within the numerous bacteria associated with insects the Pseudomonadaceae and Enterobacteriaceae families include the most important entomopathogenic species. The genera *Pseudomonas (P. aeruginosa, P. fluorescens, P. putida), Proteus (P. vulgaris, P. mirabilis), Enterobacter (E. cloacae, E. aerogenes), Serratia (S. marcescens, S. entomophila, S. liquefaciens)* among others. Many of these are a part of the insect's intestinal microbiota, however, physiological stress and mechanical, or other damages in the digestive tract of the hosts, can facilitate profuse replication in these bacterial species and cause insect bacterioses. Physiological stress is usually linked with the insect's diet and/or morphological damages caused by microparasite activity such as microsporidia, gregarines and nematodes. The penetration of bacteria into the insect's body cavity leads to acute septicemia when the bacteria disperse and multiply in the insect's hemolymph. Some species of bacteria possess the ability to penetrate into the insect body cavity in the period of metamorphosis, when a new digestive tract is formed. Metamorphosis is the critical life stage in both natural and laboratory insect populations.

2.3 Rickettsial Diseases

Entomopathogenic rickettsiae are grouped in the family of Rickettsiaceae. They are obligate intracellular pathogens, very similar to bacteria. These microorganisms present a strong association with arthropods, as symbionts or pathogens. The principal entomopathogenic species are found in the genus *Rickettsiella*.

2.3.1 Morphology of the Pathogens

Rickettsiae are polymorphic gramnegative microrganisms. The size of rickettsiae ranges from 0.2 μm to 0.3 μm in width and from 0.3 μm to 3.0 μm in length. Usually rickettsiae present small straight or bent rod morphology, and sometimes a spherical shape. The pathogenic forms produce crystals with 0.5–1.0 μm in insects.

2.3.2 External Signs and Symptoms of Disease

Symptoms of rickettsioses in the scarab beetle *Melolontha melolontha* show a milky color due to the accumulation of a large number of rickettsiae in the insect's body followed by a lost in the insect' turgor.

2.3.3 Pathomorphology of Disease

Rickettsiae predominantly infect the cells of the fat body and hemocytes, however, in the acute phase of infection the microorganism can be observed in the tracheal matrix, the Malpigian tubes, ganglions, ovaries cells and the intestine epithelium. The initial sign of disease is the appearance of big groups of rickettsiae in the form of spots. The zone of agglomeration of microbial cells enlarges and broadens in the fat body. The infected cells are strongly vacuolated. Ultimately the cells are destroyed and rickettsiae reach the hemolymph.

2.3.4 Hosts

Rickettsioses are known to infect species of Coleoptera (*Rickettsiella melolontae, R. tenebriones, R. popilliae, R. stethorae*), Diptera (*R. chironomi, R. tipulae*) and Orthoptera (*R. grylli, R. schistocercae*).

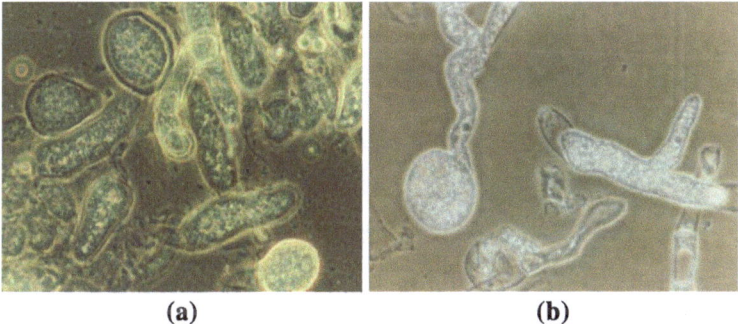

(a)   **(b)**

Fig. 2.23 Morphological structures of Entomophthoralean fungus *Entomophthora sp.*. Hyphal bodies. Phase contrast, objective 40x

2.4 Fungal Diseases

Entomopathogenic fungi comprise a vast group of microorganisms distributed in a wide number of taxa, but the most important insect pathogens are found in the class Zygomycetes and Hyphomycetes. The Zygomycetes include the order Entomophthorales which comprises obligatory insect pathogenic fungi. Entomophthoralean fungi are very important factors regulating natural insect populations. These pathogens can also cause insect mortality in laboratory. Hyphomycetes fungi cause different types of muscardine mycoses.

2.4.1 Entomophthoralean Diseases

2.4.1.1 Morphology of the Pathogens

The main morphological structures of Entomophthoralean fungi are hyphal bodies (Fig. 2.23), the zygospores (or resting spores) and different types of conidia. Usually, the resting spores present rounded or ovoid forms with thick walls (Figs. 2.24, 2.25). Spore diameter ranges from 20 μm to 40 μm and sometimes significantly more. The conidia are variable, and can be rounded, pyriform, and ellipsoidal (Fig. 2.26). The size of germinated conidia varies broadly (Figs. 2.27, 2.28, 2.29). Resting spores and conidia are the most important diagnostic structures.

2.4.1.2 External Signs and Symptoms of Disease

External manifestations of entomophthoralean diseases show a broad array of signs. Some infected insect species perch on the upper part of plants or any other

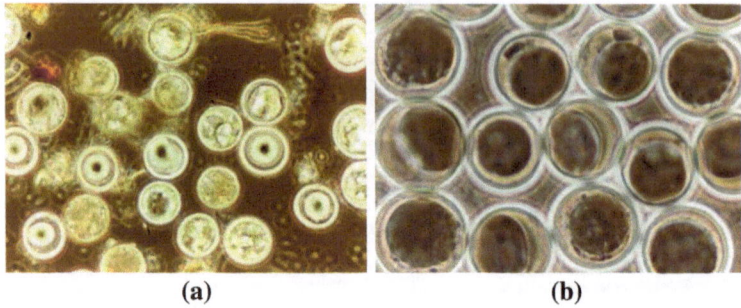

Fig. 2.24 Morphological structures of Entomophthoralean fungus *Entomophaga maimaga* from gypsy moth, *Lymantria dispar*. Rounded forms of resting spores. Phase contrast **a** Objective 40x; **b** Objective 90x

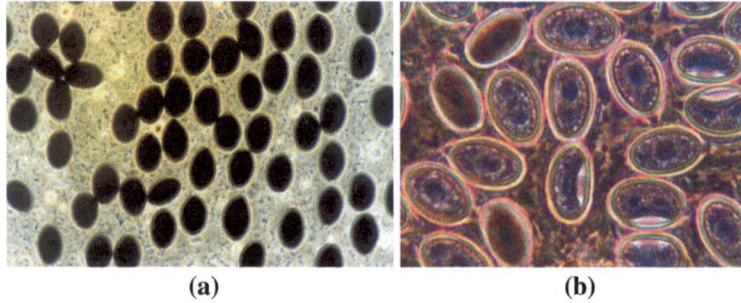

Fig. 2.25 Morphological structures of Entomophthoralean fungus *Neozygites sp.* From hemolymph of aphid *Dactinotus nigrotuberculatus*. Ovoid forms of resting spores. Phase contrast **a** Objective 40x; **b** Objective 90x

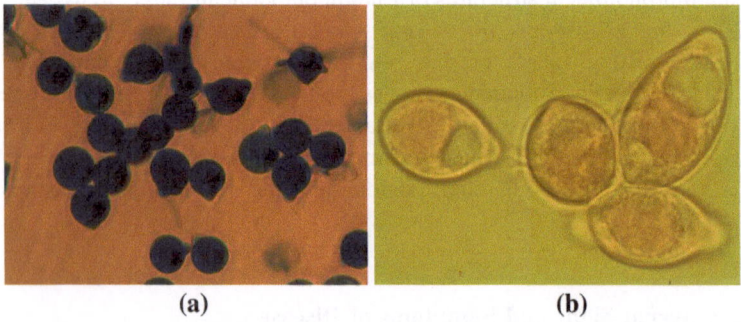

Fig. 2.26 Morphological structures of Entomophthoralean fungi. Conidia. **a** Fungus *Conidiobolus sp..* Cotton blue stain, phase contrast objective 40x; **b** Fungus *Entomophaga sp..* Eosin stain, phase contrast, objective 90x

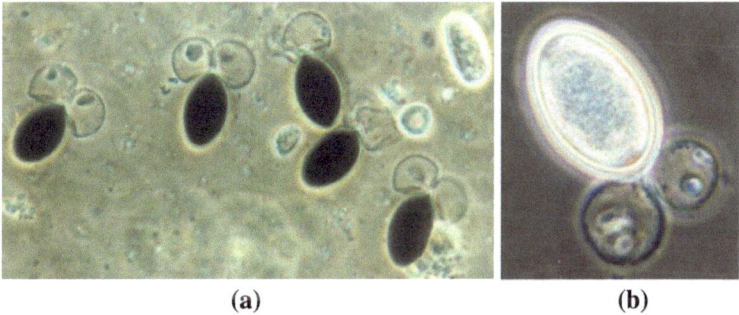

(a)   (b)

Fig. 2.27 Morphological structures of Entomophthoralean fungus *Neozygites sp.* from hemolymph of aphid *Dactinotus nigrotuberculatus*. Germinated resting spores. Phase contrast **a** Objective 40x; **b** Objective 90x

Fig. 2.28 Morphological structures of Entomophthoralean fungus *Neozygites sp.* Germinated conidia. Cotton blue stain, phase contrast, objective 90x

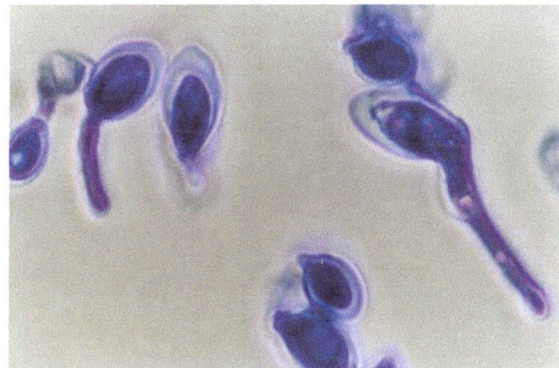

Fig. 2.29 Morphological structures of Entomophthoralean fungus *Zoophthora sp.* From aphid. Germinated conidia. Cotton blue stain, phase contrast, objective 40x

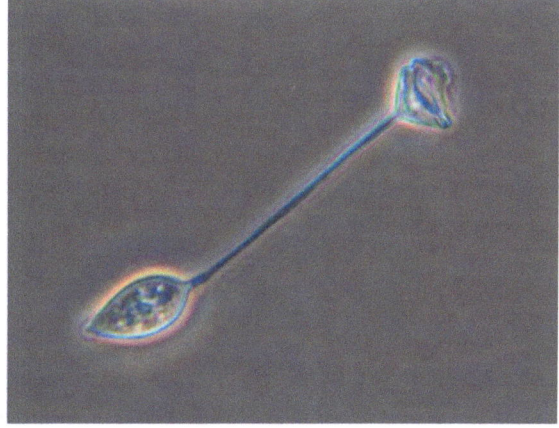

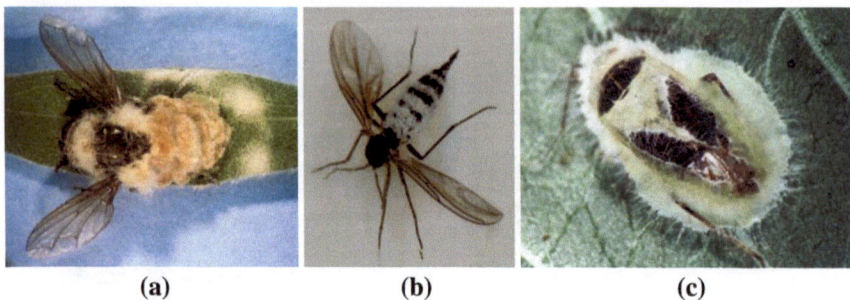

| (a) | (b) | (c) |

Fig. 2.30 Manifestation of Entomophthoralean disease in Diptera (**a**, **b**) and Heteroptera (**c**) insect

Fig. 2.31 Locust infected by entomophthoralean fungus *Entomophthora grylli*

high substrata. The initial stage of diseases is characterized by a decrease in the insect's activity level, and later by the fungal coating of the entire insect body. The coating can be very profuse (Fig. 2.30) or hardly visible (Fig. 2.31). Dead insects become mummified in a short period of time.

2.4.1.3 Pathomorphology of Disease

Entomophthoralean fungi cause systemic infections. As a rule, the fungus penetrates into the insect's body through the cuticle. Initial infection starts with resting

Fig. 2.32 Incomplete phagocytosis of hyphal bodies of the fungus *Neozygites fresenii*: azure-eosin stain, objective 100x

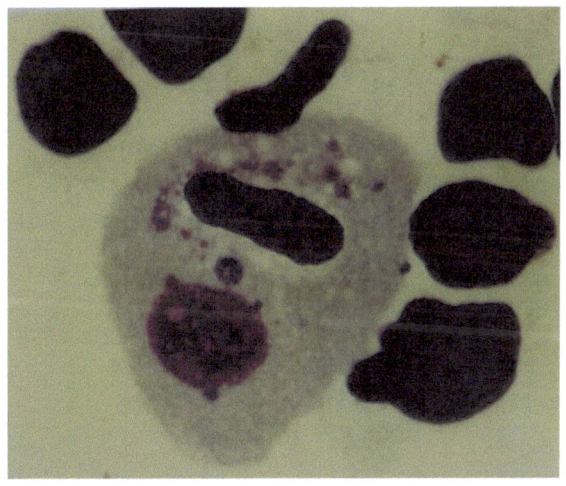

spores, which form a germ tube capable of producing enzymes allowing the fungus to penetrate into the host's cuticle. Fungal propagules quickly disperse in the insect's hemolymph and the host's organs. At this stage of infection it is possible to observe the fungal structures inside of hemocytes as a result of incomplete phagocytosis (Fig. 2.32). Resting spores and conidia are simultaneously formed in the insect's body. The mass production of resting spores and conidia takes place in the final stage of mycosis.

2.4.1.4 Hosts

Entomophthoralean fungi have a worldwide distribution and entomophthorous infections have been identified in a wide number of arthropods including ento-mophagous species.

2.4.2 Muscardine Diseases

Fungi from the class Hyphomycetes cause muscardine diseases. The most relevant genera of entomopathogenic fungi include *Beauveria, Lecanicillium, Metarhizium, Paecilomyces,* and *Sorosporella*. All these fungi develop colorless or pigmented mycelium. Conidiophores are directly formed in mycelia and can present simple or ramified structures. Conidia can present different shapes: spherical, oval, ovoid, elliptical, etc.

2.4.2.1 Morphology of the Pathogens

The range of colors in muscardine diseases comprises white, green, pink and red pigmentation (Figs. 2.33, 2.34).

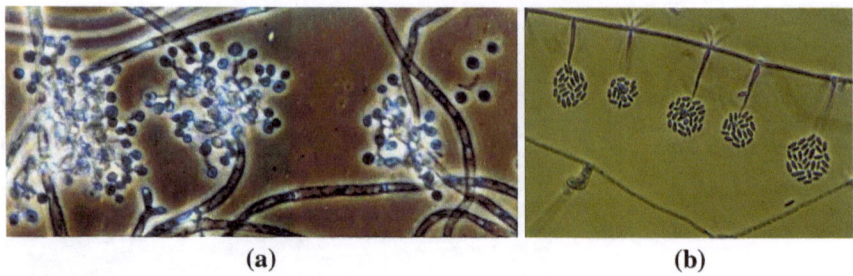

Fig. 2.33 Entomopatogenic hyphomycetous fungi. **a** *Beauveria bassiana*, cotton blue stain, phase contrast, objective 100x; **b** *Lecanicillium muscarium*, cotton blue stain, phase contrast, objective 40x

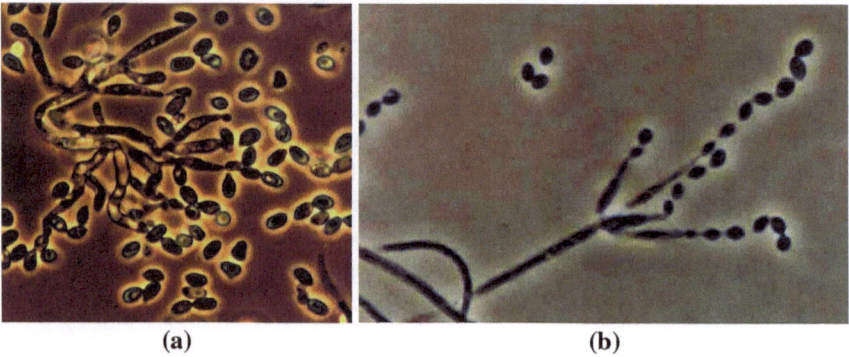

Fig. 2.34 Entomopathogenic fungi. **a** *Metarhizium anisopliae*. **b** *Paecilomyces fumosoroseus*. Cotton blue stain, phase contrast, objective 100x

White muscardine disease is caused by fungi in the genus *Beauveria*. The most important species is *B. bassiana*. The fungus produces numerous spherical and oval conidia from 1 to 4 μm in diameter. Hyphae of *B. bassiana* are septated with 3.5 μm in diameter (Fig. 2.33a). *Lecanicillium muscarium* also exhibits white propagules. It is an important insect pathogen. The fungus forms thin hyphae, 1.2–2.5 μm in diameter, septate, and powder-like after sporulation. Conidiophores range from 40 μm to 110 μm long and 1.4–2.5 μm wide with ovoid conidia located in globose clusters (Fig. 2.33b). The conidia have sizes from 2.5 μm to 6.2 μm long and from 1.0 μm to 2.5 μm wide.

Green muscardine diseases are caused by fungi from the genus *Metarhizium*. The principal insect pathogen from this genus is the fungus *M. anisopliae*. The fungus produces green cylindrical conidia with rounded ends (Fig. 2.34a). Size of conidia ranges is approx. 4.8 μm in width and 1.6 μm in length.

Pink muscardine disease is associated with the genus *Paecilomyces*. The most well-known entomopathogenic fungus from this genus is *P. fumoso-roseus* (Fig. 2.34b). Cylindrical conidia with rounded or sharp ends form beadlike chains

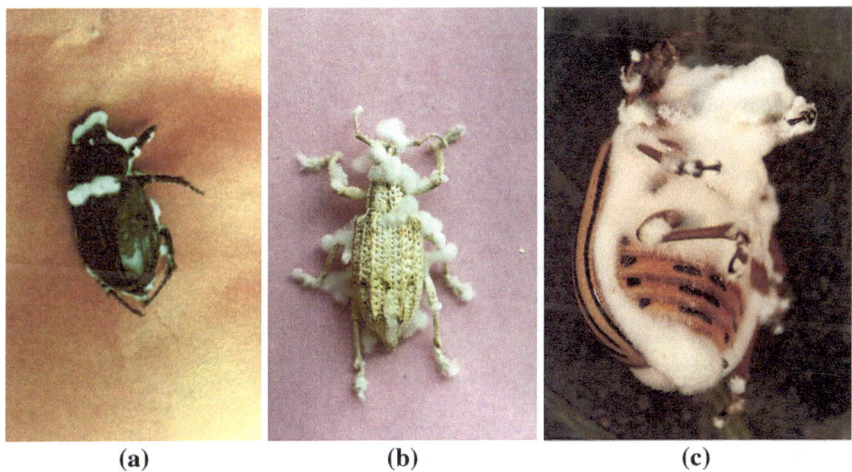

(a)   (b)   (c)

Fig. 2.35 Manifestation of white muscardine caused by the fungus *B. bassiana*. **a** Scarabaeids beetle; **b** Curculio beetle; **c** Colorado potato beetle

40–50 μm long. Sometimes the extended chain can reach 90 μm. Size of conidia range from 3–4 μm × 1–2 μm.

Red muscardine disease is caused by the fungus *Sorosporella uvella*. The fungus produces blastocysts having irregular forms with a size of approx. 8.5 μm in diameter. Oval conidia present a size of 5.3–8.8 μm × 2.6–3.5 μm. The blastocysts can form multicellular colonies with a size ranging from 52 μm to 210 μm. Some cells are able to transform into red oidia with a size of 3.5–12.3 μm.

2.4.2.2 External Signs and Symptoms of Disease

Insects infected with hyphomycetous fungi have typical external sign of diseases. The insect looses activity, ceases nutrition and dies. The insect's body is covered with mycelium (Figs. 2.35, 2.36, 2.37). Sometimes the mycelium attaches the insect's cadaver to the substratum.

2.4.2.3 Pathomorphology of Disease

Infection occurs by contact of fungal conidia spores with the insect's cuticle. Conidia than form a germ tube which penetrates through the insect's integument. Fungal cells quickly develop and form numerous blastospores. The blastospores completely fill the insect's body. The infected insects die quickly. Cadavers are covered with sporulating mycelia. Initially, fungal mycelia can be observed on the intersegmental membranes of the host (Fig. 2.34a, b) and then gradually

(a)   (b)

Fig. 2.36 Manifestation of muscardine disease caused by the fungus *B. bassiana* and *M. anisopliae* in brown marmorated stink bug, *Halyomorpha halys*. **a** White muscardine; **b** Green muscardine

Fig. 2.37 Mycosis of hymenoptera insect caused by fungus *P. fumoso-roseus*

progresses to the insect's body, ultimately enclosing the complete body of the insect (Fig. 2.34c).

2.4.2.4 Hosts

Hyphomycetous fungi cause diseases in all groups of insects in different climatic zones around the world.

2.4.3 Microsporidia Diseases

Microsporidia are unicellular parasites linked to different groups of insects and other animals, including mammals. The systematic position of the Microsporidia is

Fig. 2.38 Spores of *Pleistophora schubergi* from brown tail moth, *Euproctis chrysorrhoea*. Scanning electron microscopy, 15,000x

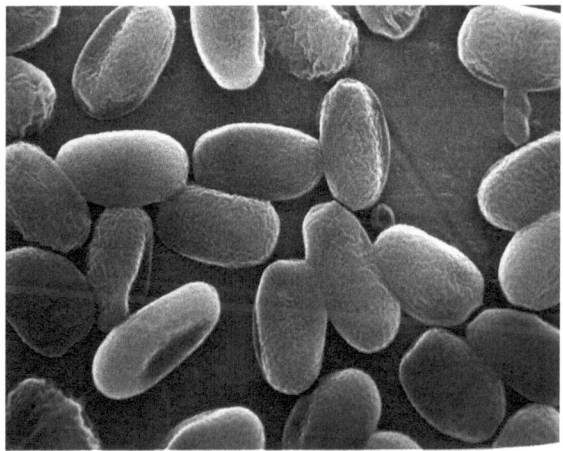

under discussion. Traditionally this group of pathogens was included in the kingdom Protozoa as an independent phylum, Mycrospora. However, at present time, microsporidia are included in the class Zygomycetes, kingdom Fungi. Microsporidian parasites are the most important pathogens influencing the health and viability of insects in laboratory or industrial cultures. As a rule, diseases caused by microsporidia do not manifest with any clear visible symptoms. However, microsporidia have influence in the behavior, fertility, metamorphosis, and diapauses of insect hosts. Many hundred species of microsporidia have been recorded in insects, predominantly in the order Lepidoptera, Diptera, Coleoptera, and Orthoptera. New species of this pathogen are routinely described. Species in the genera *Nosema, Pleistophora* and *Thelohania* are widely distributed in insect populations.

2.4.3.1 Morphology of the Pathogens

The final stadium of microsporidian development is the spore. Spores present an identical form for all species. Usually they are oval, ovoidal, cylindrical with rounded ends or pyriform (Figs. 2.38, 2.39). The size of spores varies from 3 μm to 6 μm long and from 2 μm to 4 μm width, seldom larger. The spores can be singly located or can be in a group (sporonts). Sporonts are a characteristic feature of microsporidia. For example, species in the genus *Nosema* produce spores distributed in sporonts; species in the genus *Pleistophora* present ≥16 spores per sporont whereas the species in the genus *Thelohania* present 8 spores per sporont. Microsporidia spores show a complex morphological structure. The spore wall has three layers including external proteinaceous, middle chitinous, and internal plasmatic membrane (Fig. 2.40). The spore structure is adapted for penetration into host cells through a special morphological structure, the hollow polar tube (polar filament). The polar filament significantly exceeds the size of the

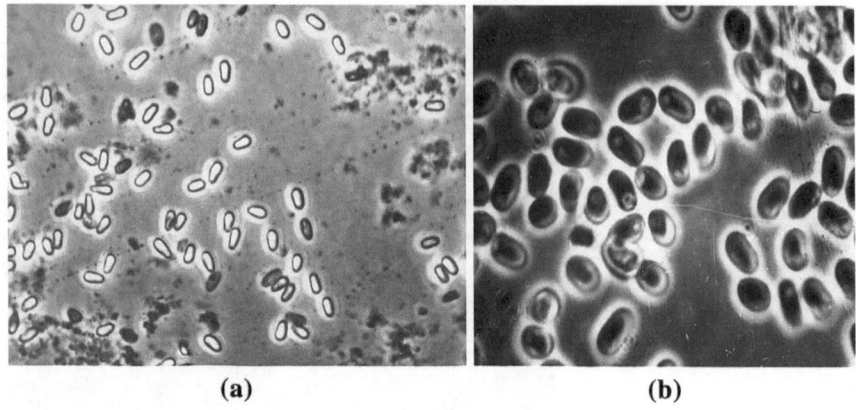

(a)   (b)

Fig. 2.39 a Spores of *Nosema sp..* , light microscope, phase contrast, objective 40x; **b** Spores of *Thelohania sp..* objective 90x

Fig. 2.40 *Nosema* spores
with visible layers of the
spore wall. Scanning electron
microscopy, 80,000x

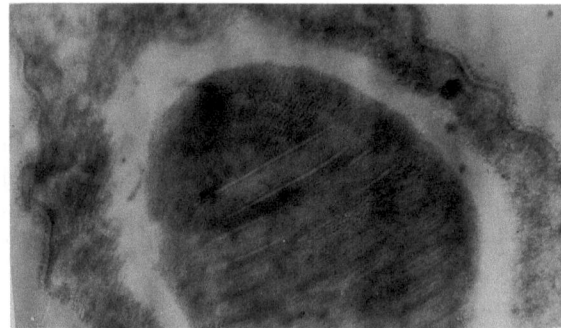

spore with relation to the base, located inside the spore, and arranged in a spiral (Figs. 2.41, 2.42).

The microsporidia have a life cycle consisting in two phases, the merogony and the sporogony. The merogony is accompanied by numerous replications of the parasite with formation of meronts that can be uninucleate, binucleate, tetra nucleate, octonucleate, and multinucleate. The meronts present different morphologies and sizes. The sporogonial stage comprises the sporont, which after multiplication forms a sporoblast in which mature spores are formed.

2.4.3.2 External Signs and Symptoms of Disease

Signs or symptoms of microsporidian infections are not easily visible in insects with pigmented integument. Symptoms of infection are easier to observe in insect lacking pigmentation (Fig. 2.43). Pigmented insects can present unusual coloration. For example, gypsy moth and fall webworm butterflies infected by

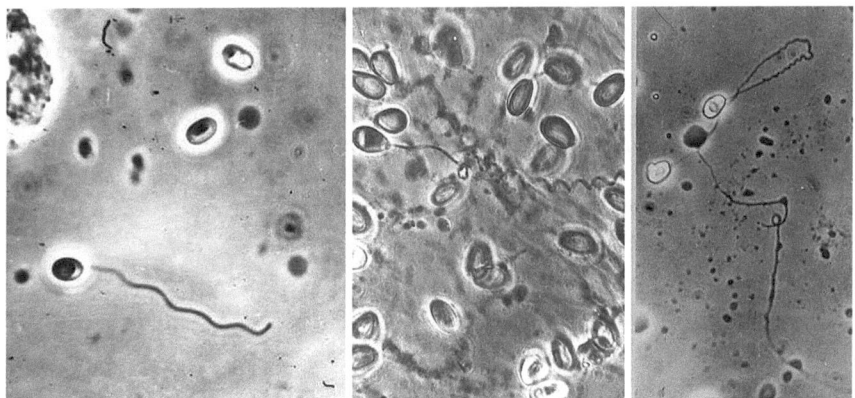

Fig. 2.41 Spores of microsporidia with polar filament. Phase contrast, objective 40x

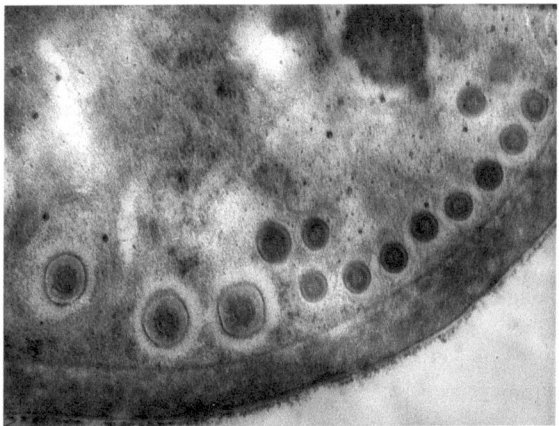

Fig. 2.42 Transversal section of *Nosema sp.* spore showing the spiral structure of polar filament. Scanning electron microscopy, 1,20,000x

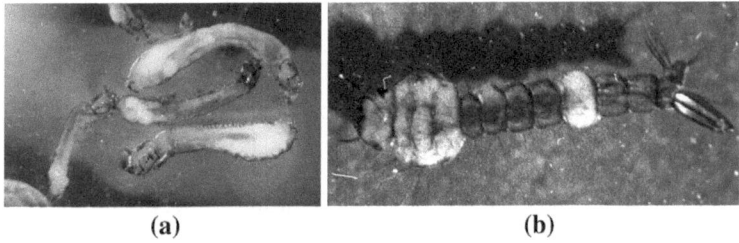

(a)   **(b)**

Fig. 2.43 a Unpigmented larvae of midges; **b** Mosquito with visible sign of microsporidia infections

Nosema spp. microsporidian show yellow pigmented bodies. The disease insects can often have black spots on the body. Irregular growing patterns are observed in larvae with the same age. This peculiarity is especially visible in laboratory populations of insects. Lepidoptera larvae have noticeable disproportions between the head and body. Infections produced in the gut and Malpigian tubes loose their water balance. Water is excreted from the gut, and the insect's organism becomes dehydrated. As a result, the insect's body wrinkles and darkens.

2.4.3.3 Pathomorphology of Diseases

Microsporidia present tissue tropisms. Species cause generalized infections on different types of tissues of the host. For example, the microsporidia *Nosema bombyces* infects different organs and tissues of silkworm, *Bombyx mori*, whereas *Thelohania* spp. more often developed in the fat body or the muscle. Parasites from the genus *Pleistophora* attack mainly muscular tissue. Microsporidia can infect the insects' tissues only in the definite period of development. In populations of insects reared in laboratory the most important microsporidian infections are *Nosema* species. As a rule, these parasites cause generalized infection with destruction of all organs and tissues. Initially, the insects are infected through contaminated food. Epithelial cells of the midgut can be contaminated and destroyed. Than, the parasite penetrates through the gut wall into the hemolymph and progresses into different insect organs (Figs. 2.44, 2.45, 2.46). The spores are initially activated by influence of digestive enzymes. High pressure is generated inside the spore which can lead to the evagination of the polar filament producing a shot like appendage. The polar filament destroys the cell wall and the parasite penetrates into the cytoplasm. Practically all microsporidia develop in the cytoplasm of a cell. The parasitized insect cells start to be destroyed when parasites reach the sporogony phase.

2.4.3.4 Hosts

Microsporidia are widely distributed in all groups of insects around the world. Around 90% of all known microsporidian diseases of terrestrial insects are caused by parasites in the genera *Nosema*.

2.5 Diseases of Protozoan Origin

According to modern classification, protozoans causing insect diseases are included in the phylum of Alveolata. The phylum of Alveolata includes several subphylums (Dinoflagellata, Rhizopoda, Apicomplexa and Ciliophora) which comprise numerous species with pathogenic activity towards insects. However, species with

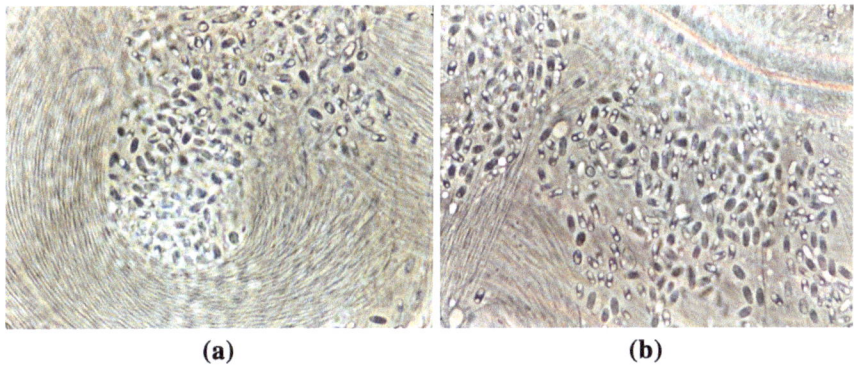

(a)   **(b)**

Fig. 2.44 a Localized and **b** General distribution of microsporidia spores in muscular tissue of *Trirhabda canadiensis* beetle. Phase contrast, objective 40x

Fig. 2.45 Sporoblasts of microsporidia *Thelohanis sp.* in insect's fat body. Phase contrast, objective 70x

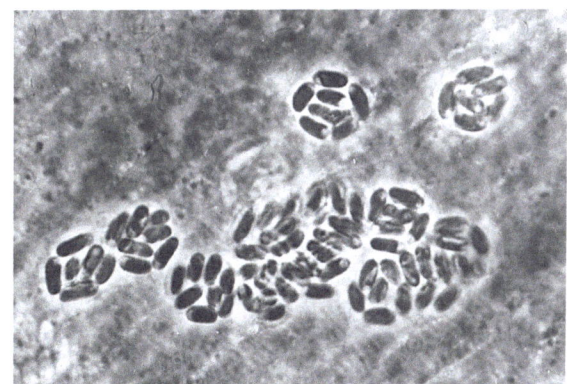

Fig. 2.46 Sporoblasts of microsporidia *Thelohanis sp.* in insect's blood. Phase contrast, objective 70x

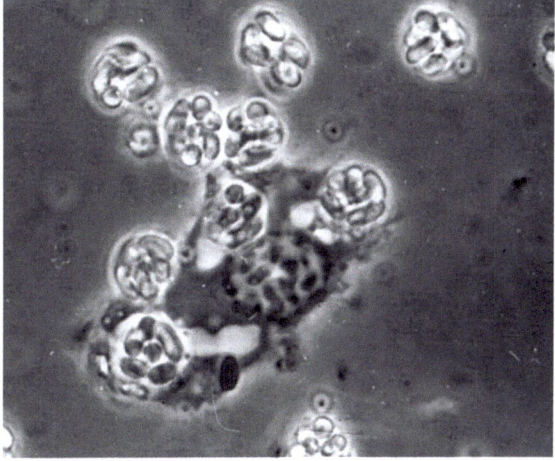

importance in laboratory and industrial insect cultures are grouped in a few species in the phylum of Rhizopoda and Apicomplexa. Protozoan species from the phylum of Rhizopoda cause amebiases disease and species from the phylum of Apicomplexa cause gregarinoses diseases.

2.5.1 Amebiases of Insects

The phylum of Rhizopoda contains two families with entomopathogenic species, the Amoebidae and Endamoebidae. The first family has three important entomopathogenic genera, the *Malamoeba, Malpighamoeba* and *Malpighiella*. Most of the amoebae species are commensals, dwelling into the digestive tract of insects. The amoebiasis disease caused by the parasite *Malpighamoeba mellificae* is a well-known problem in honey bees. Amoebiases are also found in cockroaches, *Blatta orientalis*, and several species of grasshoppers, among others. The best known species is *Malpighamoeba locustae* parasiting various species of grasshoppers of the genus *Melanoplus* (*M. mexicanus, M. femur-rubrum,* and *M. differencialis*).

2.5.1.1 Morphology of the Pathogens

Parasites produce uninuclear oval cysts which can survive for years in the environmental. The cysts of amoebae species are oval or slightly elongated with sizes ranging from 8.5 μm to 10 μm long and from 4.6 μm to 6.2 μm wide. As a rule, the cysts penetrate in the insect's body through ingestion. Cysts in the digestive tract form primary trophozoites, which penetrate to epithelial cells of the midgut and cells of the Malpighian tubules. Different species of amoebas have variable size of trophozoites ranging from 4 μm to 50 μm in diameter. Primary trophozoites can generate secondary trophozoites.

2.5.1.2 External Signs and Symptoms of Disease

External symptoms of amoebiases depend on the rate of infection. Intense infections are associated with less activity in the host and loss in appetite. Insects then become sluggish and are not able to keep a normal vertical position. In a final stage death overcomes.

2.5.1.3 Pathomorphology of Disease

Severe infections are accompanied by pathological changes in Malpigian tubules. Big masses of cysts block tubules, which can be destroyed, and as a result, allowing the parasite to penetrate into the hemolymph. The infected tubes regenerate and form spherical indurations. Parasites also block epithelial secretory cells and elimination of residues through urates is inhibited.

2.5.1.4 Hosts

The amoebiases are present in different geographical zones around the world. As a rule the hosts of entomopathogenic amoebas are insects having an economical or epidemiological interest to man. The most well known amoebas have been studied in the honey bee (amoebiases caused by *Malpighamaeba mellificae*), and different species of cockroaches (amoebiases caused by *Endamoeba blattae*, *Endamoeba thomsoni* and *Endolimax blattae*).

2.5.2 Gregarinoses of Insects

The phylum of Apicomplexa includes two classes having importance as insect parasites: Gregarinia and Coccidia. Traditionally, the phylum of Apicomplexa includes the class Microspora or Microsporea, however, at present time this large group of insect pathogens is allocated in several orders in the class Zygomycetes (kingdom Fungi). Species related to the class Gregarinia are divided into two groups, Eugregarinida and Neogregarinida, according to their reproductive cycle.

2.5.2.1 Morphology of the Pathogens

The principal morphological stages of the Eugregarinida include cysts, sporozoites and gamonts (synonym to trophozoites). The Neogregarinida species have additional morphological schizonts. The most part of mature gregarines (gamonts or trophozoites) are large, with approx. 10 mm in length and visible to the naked eye (Fig. 2.47). The parasite allocates extracellularly. As a rule, the mature parasites are found in the digestive tract. Insect excrements are a pathway for release of the parasite which then completes its life cycle forming cysts to survive until a suitable host ingests them. The cysts have an oval form, with approx. 300 μm in diameter. Cysts can live long, surviving in harsh environmental conditions. This stage is responsible for new infections and proliferation in insect colonies.

Generally the gregarines parasitize only the digestive tract of insects, but some species can penetrate into the insect's body cavity. In the latter, insects can be killed. The most part of gregarine species cause chronic insect infections, but these infections can stimulate septicemia through enhanced activity of semi-saprophytic and saprophytic microorganisms permanently living in the digestive tract of insects. Gregarinoses can also be responsible for a decline in fecundity and egg viability.

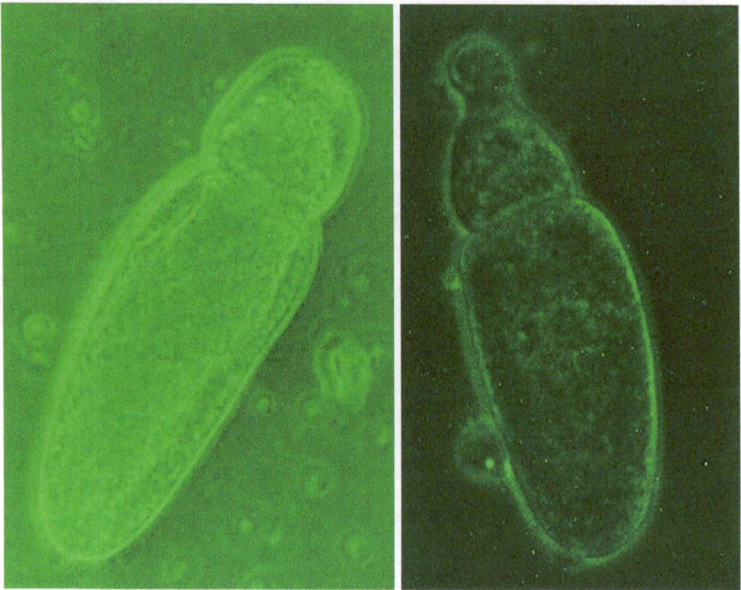

Fig. 2.47 Gregarine in *Trirhabda canadiensis* beetle. Phase contrast, objective 10x

2.5.2.2 External Sign of Disease

Insects infested with gregarines do not have visible symptoms of infection. Heavy infections can cause diarrhea. Infested insects also show nonspecific disease signs including decline in nutrition, drop in weight and limpness.

2.5.2.3 Pathomorphology of Disease

In most cases gregarines are located in the digestive tract and Malpighian tubules. Parasites damage the midgut epithelium and cells of the Malpighian tubules.

2.5.2.4 Hosts

At present time most species of gregarines associated with insects are know in the Coleoptera, Diptera and Lepidoptera insect orders.

2.5.3 Coccidioses of Insects

Coccidians are intracellular parasites. This group of microorganisms is related to different groups of invertebrate and vertebrate animals. Some species have very important significance in fish farm ponds. The most important insect pathogens are found in the genus *Adelina*.

2.5.3.1 Morphology of the Pathogens

The life cycle of coccidian parasites is complex and includes merogony, gamogony and sporogony phases. This life cycle is very efficient for a successful and productive reproduction of the parasites. The main morphological stages of the Coccidians include cysts (oocysts), spores, sporozoites, meronts, merozoites, schizonts, gametocytes, sporonts and sporoblasts. The size of the cysts depends on the number of spores (sporocysts). For example, cysts containing five spores have 26 μm in diameter and cysts with 21 spores range from 46 μm to 51 μm. Sporozoites and gametocytes have vermicular forms that vary in size. Usually merozoites present a banana-shaped form with 8 μm to 16 μm in length. The schizonts are multinucleate bodies and these morphological features can be easily discernible by light microscopy.

2.5.3.2 External Sign of Disease

Insects infested with coccidian do not manifest external symptoms of infection. Heavy infections in insects show nonspecific signs of diseases including decline in nutrition, drop in weight and limpness.

2.5.3.3 Pathomorphology of Disease

The initial stage of coccidian infections is related with the presence of cysts which survive in the environment and reach the digestive tract of the insect through food ingestion. Spores (sporozoites) located inside cysts are then released and penetrate into the gut epithelium or body cavity of the host. Often, the coccidian development takes place in the fat body of the insect. All coccidians develop in a parasitophorous vacuole. In most cases the vacuoles are located inside the cytoplasm of the cells. The infected cells are hypertrophied and contain large numbers of oocysts.

2.5.3.4 Hosts

Most of the insects susceptible to coccidian parasites are in the orders Coleoptera, Lepidoptera, Orthoptera and Diptera.

2.6 Mix Infections

Quite often insect diseases are caused by obligatory microbial pathogens including, first of all, viruses, entomophthoralean fungi, microsporidia and protozoa. It is a very usual situation to have two or more species of obligatory entomopathogenic microorganisms developing in one host. Increasingly immune-compromised insects have no barriers for pathogen infection. There are numerous examples of insect diseases caused by a complex of viruses. Viral mix infections have been described in different species of insects in the orders Lepidoptera (families Noctuidae and Lymantriidae).

References

Aruga H, Tanada Y (eds) (1971) The cytoplasmic-polyhedrosis virus of the silkworm. University of Tokyo Press, Tokyo
Reardon R, Podgwaite JP, Zerillo RT (1996) GYPCHEK-The gypsy moth nucleopolyhedrosis virus product. USDA Forest Service publication FHTET pp 96–16
Tanada Y, Kaya H (1992) Insect pathology. Academic Press, NY

Chapter 3
Manifestation of Infectious Diseases in Insect Cultures

Abstract Describes the peculiarities of initial disease manifestation (cryptic, mild or acute) in insect cultures for prompt detection of abnormalities in the colony, namely external and internal symptoms and signs of disease (e.g. change of morphology and metamorphosis patterns, behavioral changes, etc.). This chapter is mostly useful for timely detection of infectious diseases in insect laboratory cultures and should be used when problems in the colonies are observed and as a regular checklist for colony health.

Keywords Timely detection · Cryptic pathogen · Mild pathogen · Acute pathogen · Colony health checklist

Infection diseases in laboratory reared insect cultures can manifest both as acute (explosive) spreading epizootics with high infection and mortality rates, or slowly developing infections (chronic) with small mortality rates. There are different transitional forms of epizootics from acute to chronic patterns of infection. Usually, acute epizootics are associated with semi-saprophytic bacteria and baculoviruses, as well as, reoviruses. Slow developing infective diseases are typically caused by microsporidia, gregarines and coccidian. For a timely assessment of the insect's condition, it is first of all necessary to pay attention to general external visible signs of diseases. These signs include:

1. Decrease in nutrition intensity by the insect;
2. Decrease of activity;
3. Change in behavior;
4. Diarrhea;
5. Changes in body turgor;
6. Changes in the external integument;
7. Appearance of unusual smells (usually a putrefaction sense);
8. Limping.

V. Gouli et al., *Common Infectious Diseases of Insects in Culture*,
SpringerBriefs in Animal Sciences, DOI: 10.1007/978-94-007-1890-6_3,
© Vladimir Gouli 2011

As a rule, acute infection diseases develop very quickly and laboratory insect populations perish due to a single of complex pathogenic activity. Laboratory and industrial populations of different cutworm species including cabbage looper, *Mamestra brassicae*, noctuid on cotton, *Heliothis armigera* and black-c owlet moth, *Graphiphora c-nigrum*, when infected with acute cytoplasmatic polyhedroses, can perish after a few days following the notice of the first disease signs of disease. Slowly developing insect diseases, with low mortality rates, can remain unnoticed for a long time. Such types of diseases are characteristic in parasitic microsporidia, gregarines and other protozoa. Insects infected with mild obligatory pathogens are then susceptible to facultative pathogens.

In an initial period of development of the infection processes, decrease of nutrition intensity and activity are, as a rule, typically for all infective diseases of insects. Behavioral reactions can have specific characteristics for each group of diseases. Insects infected with baculoviruses or Entomophthoralean fungi can move to the upper side of rearing cages or plans. Soil-dwelling beetle larvae infected with rickettsia do not try to seek shelter and move to open spots.

Diarrhea is a very important sign of insect diseases. Different types of diseases are accompanied by dysfunction in digestion processes. Acute diarrheas can usually be observed in insect bacterioses. As a rule, intestinal excretions have black color and putrefied smell. Same bacteria are able to generate pigments which can influence the color of the excrements and integuments. For example, the facultative insect pathogenic bacterium *Serratia marcescens* produces red pigments and bacterioses caused by this microorganism are accompanied the diarrhea with red color.

Insect body turgor changes depending on pathogen type and infection stage. Obligatory insect pathogens developing in the fat body cause hypertrophy of cells and, as a result, the body turgor of the host insect increases. However, saprophytic and semi-saprophytic bacteria produce enzymes which destroy the cell's membrane, and the body turgor of the host insect quickly decreases. The insect external integument changes color and tensile strength. Unfortunately this typical external signs of diseases are observed not so often because bacteria in commensal host-insect relationships rapidly multiply in presence of a physiological stressful situation and cause the darkening of insect's body. Proliferations of bacteria are a reason for the unusual smell present in insects with bacterial infective diseases.

Chapter 4
Methods of Practical Diagnostic

Abstract Includes detailed guidelines for practical, expedient and accessible diagnostic methodologies of infectious diseases in insects. Techniques described provide reliable alternatives for preliminary pathogen identification which do not require special equipment or intensive training, in a step by step protocol format, and for an easy orientation and accurate learning.

Keywords Practical methodologies · Pathogen identification · Step by step protocols

4.1 Preliminary Diagnosis

Preliminary diagnosis includes the assessment of specific external signs of disease. In the case of appearance of any external signs of disease it is necessary to conduct preliminary simple microscopic analysis of the excrements and the hemolymph to determine the causal agents. Numerous insect pathogens parasitizing the gut of the host and the resting stages of parasites can be observed in the excrements. Sometimes the simple microscopic analysis using a "squashing drop method" can led to the identification of the insect disease. This method includes the following procedures:

1. Pick up a droplet of water on a slide.
2. Pick up a small amount of insect excrements (closely in size to the water droplet).
3. Mix up the droplet of water with excrement.
4. Cover the mixture with a cover slide and examine under a phase contrast microscope.

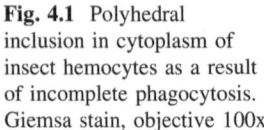

Fig. 4.1 Polyhedral
inclusion in cytoplasm of
insect hemocytes as a result
of incomplete phagocytosis.
Giemsa stain, objective 100x

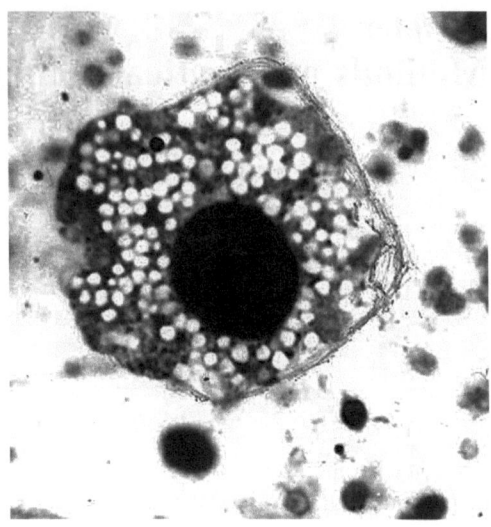

The "squashing drop method" allows recognizing the pathogens located in the insect's digestive tract including digestive polyhedroses, bacterial septicemia, microsporidia and protozoa.

Preliminary diagnosis can also include a simple examination of the insect's hemolymph. The method includes the following procedures:

1. Pick up a droplet of physiological solution (or any artificial media for insect cell cultivation) on a slide;
2. Obtain a droplet of hemolymph from the insect's body. (Note: droplets of hemolymph can be easily received by excising an insect leg or making an incision with a scalpel in the insect body);
3. Quickly allocate the droplet of hemolymph into the surface of the physiological solution. The drop of hemolymph should be placed near one end of a slide.
4. Cover the mixture with a cover slide. Carefully draw the cover slide until it contacts the drop. In this case the drop will spread across the edge of the slide, and than push the top slide for an even distribution of the hemolymph droplet.
5. Examine under phase contrast microscope.

The majority of infectious insect diseases are accompanied with intensive vacuolization of the hemolymph (Fig. 4.1). The vacuolization of hemolymph is not a specific reaction to a pathogen but this reaction allows visualizing the pathological processes involved in the infection. Sometimes it is possible to distinguish pathogens in the cytoplasm of hemocytes as a result of incomplete phagocytosis (Figs. 2.32, 4.1).

In the case of fungal infections the most effective and simple "scotch tape method" can be used (Gouli et al. 2005). Usually, mycoses of insects are associated with the appearance of different fungal propagules in the surface of the insect host. To prepare slides for scotch tape technique, fungal propagules in the insect's hypoderm should gently contact a piece of scotch tape and subsequently allocate it in a slide with a drop of cotton blue stain.

4.2 Preparation of Biological Material for Analysis

Insects with visible abnormal morphology or behavior should be subjected to a more detail analysis. There are numerous methods for disease diagnostics but for practical work it is possible to use some simple methods. After establishing that external signs of disease are present it is necessary to prepare several coats of hemolymph, fat body and middle part of the gut for staining and subsequent microscopic examination. Be aware that in order to assess differences in the insects it is very useful to always do a parallel analyses between healthy and disease insects. Comparison slides from healthy and sick insects will allow to assess the morphological peculiarities in infected individuals contrasting with healthy ones. Slides dry quickly and should be stained rapidly. Samples for preparation of slides from the insect's fat body and middle part of gut can be obtained from previously prepared insects (Fig. 2.15). Small pieces of fat body are allocated into a small volume of water, than homogenized, and dried. Identical chirurgical operations are performed for the of the middle part of the gut. All slides can be fixed using ethanol (100%) during 10–15 min. After drying the slides, they should be stained with Giemsa stain during 1–2 min, and carefully rinsed several times in distilled water.

4.3 Microscopic Examination of Insect Tissues

Slides with sample material are examined under phase contrast microscope using objective magnifications such as 40x and 100x. Hemolymph should be the primary sample where to conduct analysis since hemolymph participates in all physiological processes of the insect, thus reflecting the general insect condition (Figs. 4.2 and 4.3a, b). Since hemolymph is liquid, slide preparation does not demand complex histological methodology for microscopic analysis. A complete hematological analysis involves the establishment of a ratio between the different types of insect blood cells. However, it is possible to conduct a more simple analysis.

Analysis of the fat body is conducted in order to expose viral inclusions such as nuclear polyhedra, granules, poxvirus rhomboidal or ovoid viral formations and different stages of microsporidia. Nuclear polyhedroses are accompanied by strong hypertrophy of the cell nuclei which contains numerous viral inclusions. Under the microscope nuclear polyhedroses present an oval form and intensive refracted light. In native slides of fat body, with no stain of fixation, and for the case of granuloses, it is possible to observe cell hypertrophy and disintegration of nuclei.

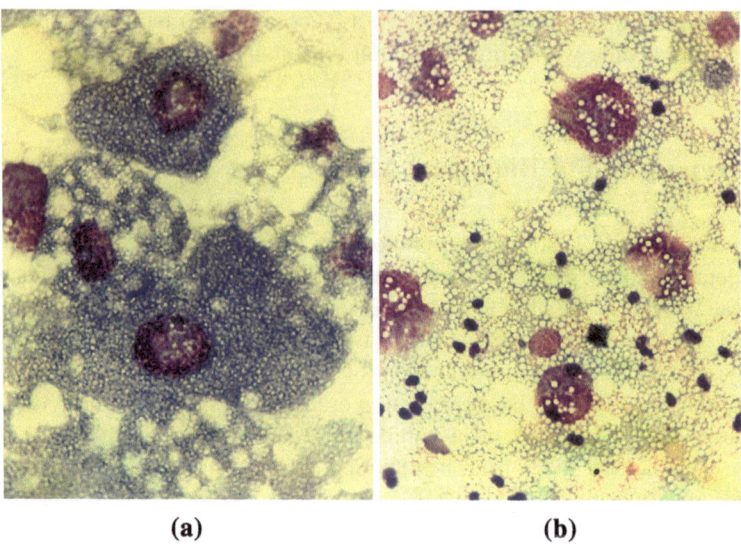

(a)   (b)

Fig. 4.2 Hemocytes of hemlock woolly adelgid, *Adelgid tsugae*. **a** Vacuolization caused by the entomopathogenic fungus *Beauveria bassiana*. **b** Vacuolization with total disintegration of cells. Giemsa stain, Light microscope, phase contrast, objective 100x

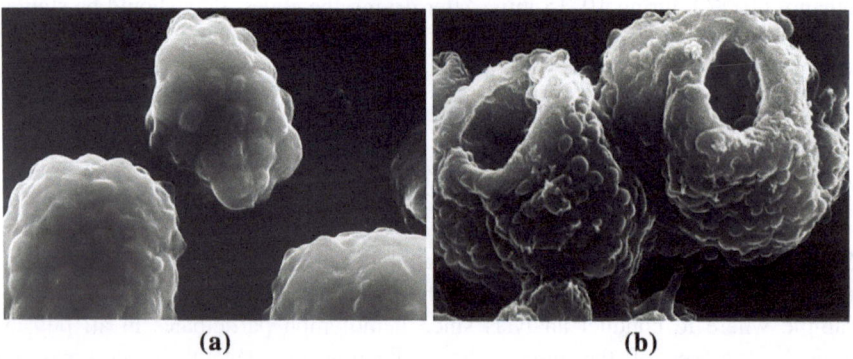

(a)   (b)

Fig. 4.3 **a** Hemocytes from healthy larvae of codling moth, *Carpocapsa pomonella*. **b** Hemocytes infected by granulosis virus with specific vacuolization. Scanning electron microscopy, 10,000x

The granules can be observed with phase contrast or darkfield microscopes. Viral inclusion of entomopoxviruses are observed in the cytoplasm of the fat body (Fig. 2.17). Slides from the middle part of the gut are generally used for diagnosis of nuclear polyhedroses of hymenopterans insects, cytoplasmic polyhedroses of Lepidoptera and microsporidioses of different groups of arthropods, especially Coleoptera.

In case of intestinal polyhedroses the epithelial cells are hypertrophied and destroyed (Fig. 2.10). Identification of cytoplasmic polyhedroses is not difficult

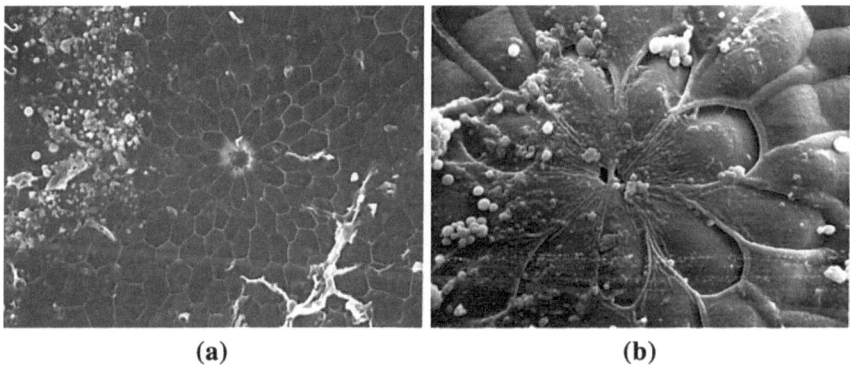

(a)   (b)

Fig. 4.4 **a** Egg micropyles of Siberian moth, *Dendrolimus superans sibiricus*. **b** Cabbage moth, *Mamestra brassicae*, with crystals of polyhedra-like uric acid salts on the chorion surface

since the cytoplasm of epithelial cells of the mid gut contains many polyhedral inclusions (with an average size of 5 μm) which are easily visible under light microscopy. Microsporidia can be observed in different insect organs and tissues but some species are only present in the middle part of the gut. Pathogenic protozoa can be observed by analyzing the content of the back part of the gut.

Diseased insect females can produce eggs containing entomopathogenic microorganisms (e.g., microsporidia or viral inclusions). The microorganisms can be located both on the surface of the chorion or under it (Fig. 4.4).

Analyses of eggs for pathogen identification is not an easy task. The simplest method is to squash the eggs directly on the slide and prepare material for microscopic analysis using wet slide mount techniques. However, as a rule, eggs contain a small number of the pathogens, and as a result, it is necessary to use a large number of eggs in the analysis. To achieve this, a group of eggs are homogenized in a 10-times volume of water, then filtered through 2–3 layers of gauze and finally the suspension obtained is centrifuged. The sediment is suspended again in water and the suspension is centrifuged once more. This operation is repeated 2–3 times. Finally the sediment is used for preparation of slides using wet slide mount techniques and an appropriate stain.

Very often, slides prepared from diseased insects contain some morphological structures that resemble pathogenic microorganisms. These structures make it difficult to determine the causal agent of disease. For a differential diagnostic, metabolites produced by viral inclusions can be useful to discern pathogens in dry material. These metabolites present specific physical and chemical properties which do not occur in the abundant crystals of uric acid salts, protein and melanin bodies are originated in the metamorphosis period of insects. Melanin crystals have a coloration ranging from yellowish to dark brown.

The identification of viral inclusions is based in methods which take into consideration the resistance of viral particles to different stains. For coloration of viral polyhedra and granules it is preliminary used a treatment of alkaline solution

(1%) during 1–2 min, with subsequent staining using any stain such as Giemsa stain. Without an alkaline treatment the color effect will be reversed, i.e., viral inclusions will lack coloration and the remaining biological material will stain. Evans and Shapiro (1993) provide an alternative to color baculovirus inclusion bodies, which does not require a preliminary alkaline treatment, using Buffalo Black 12 stain. This stain allows to distinguish crystalline proteins from other tissue materials. In this case let a tissue sample on a slide to air dry and then stain the dry smear fixed in the slide (until it is fully covered with stain) and leave in a warm environment (40–45 C for 5 min). Afterwards wash the slide under running tap water for 10 sec. The slide is now ready to be examined under light microscope with oil immersion objectives.

There are numerous methods for differential staining of major groups of entomopathogenic microorganisms (Didier et al. 1995; Humber 1997; Beveridge 2001; Noble 2001; Walochnik and Aspöck 2001), however, Giemsa stain initially developed for the identification of human blood diseases (Woronzoff-Dashkoff 2002; Barcia 2007), is the most routinely, expedient and useful stain for preliminary diagnostic of insect diseases including viruses, bacteria, fungi and protozoa.

References

Barcia JJ (2007) The Giemsa stain: its history and applications. Int J Surg Pathol 15:292–296

Beveridge TJ (2001) Use of the Gram stain in microbiology. Biotech Histochem 76:111–118

Didier ES, Orenstein JM, Aldras A, Bertucci D, Rogers LB, Janney FA (1995) Comparison of three staining methods for detecting microsporidia in fluids. J Clin Microbiol 33:3138–3145

Evans H, Shapiro M (1993) Viruses. In: Lacey L (ed) Manual of techniques in insect pathology. Academic Press, London, pp 17–54

Gouli VV, Gouli S, Costa S, Shternshis M (2005) Comparison of wash, leaf imprint and adhesive tape methods for qualifying fungal spore deposition on leaves. J Mycol Phytopathol 39(3): 99–103

Humber, R (1997) Fungi: Identification. In: Lacey LA (ed) Manual of Techniques in Insect Pathology. Academic Press, London, pp 153–185

Noble RT (2001) Enumeration of viruses. Methods in Microbiology 30:43–51

Walochnik J, Aspöck H (2001) Protozoan Pathogens: Identification. In: Encyclopedia of life sciences. John Wiley & Sons, pp 1–9

Woronzoff-Dashkoff KK (2002) The Wright-Giemsa stain: secrets revealed. Clin Lab Med 22:15–23

Chapter 5
Prophylactic of Infectious Diseases in Insect Cultures

Abstract Gives recommendations for prophylactics and control of infectious diseases in insect cultures. Sanitary measures for the establishment of axenic insect cultures using gnotobiotic insects are described. In addition, insect sanitation and therapy of disease are explained. Protocols for sterile and biosafety laboratory use, as well as insect culture handling are listed.

Keywords Establishment axenic colonies · Prophylactics of disease · Biosafety · Insect colony handling · Protocols

5.1 Selection of Animals for Establishment of Healthy Laboratory Colonies

Insects intended for rearing in laboratory conditions should be free from parasitoids and entomopathogenic microorganisms. Gnotobiotic insects (pure lines) are the best material for establishment of laboratory culture of insects. Usually, natural insect populations are infected with different microorganisms, and it is critical to select healthy individuals for future rearing. Obtaining gnotobiotic insects requires special procedures in conformity to each species (Talpalatsky et al. 1984; Gouli et al. 1999). However, there are relatively simple criteria for selection of healthy individuals in natural insect populations. These criteria include:

1. Estimate variability of insect color;
2. Estimate correlation of sexes in a population;
3. Estimate population quality according to egg samples analyses;
4. Estimate population quality according to larvae appearance;
5. Control of parasitoid infestation and infection diseases.

Many species of insects change intensity of color depending on their physiological condition. Insect populations in the beginning of an outbreak present individuals with darker coloration. It has been established that darkening indicates disturbance in physiological processes and decay in insect health conditions. Sex ratio in populations is an important index to assess the condition of insect populations. Healthy insect populations have more females and fewer males. Laboratory insect populations are characterized with these same premises.

A healthy female is one that can deposit an optimal number of eggs without any signs of chorion deformation or unusual coloration. Eggs from healthy individuals should hatch in an exactly time frame for a particular species and the new insect generation must be active and devoid of any behavior deviation from the standard assessed for the species.

5.2 Insect Sanitation and Therapy of Disease

When presence of infection in insect laboratory populations is confirmed, two different procedures can be taken into practice. The decision on what procedure to apply depends in the origin of the disease. Procedures are:

1. Complete elimination of the colony following standard sterilization procedures, and restart a new pure colony. All viral infection require the radical elimination of the insect population since at present time there is no effective means for treatment of these group of infections.
2. Use sanitary methodologies for infected populations, when possible. Bacterioses, mycoses and sometimes protozoiases, can be suppressed using chemical and physical procedures. Antibiotics are effective against insect mycoses and bacterioses. Usually, the artificial insect's nutrient substrata includes antibiotics and antiseptics for stabilization of media and to prevent insect diseases. For microsporidia it has been assessed that warming the insect's rearing cages to a temperature slightly above optimal for the colony is an effective method for management of this disease group in silkworm, *Bombyx mori*, potato tuber moth, *Phthorimaea operculella* and gypsy moth, *Lymantria dispar*. Surface treatment of insect eggs using different antiseptics (e.g. sodium chloride, boric acid, phenol and methanol) is very effective for the suppression of different types of insect pathogens. Doses and treatment times vary accordingly with each insect species.

5.3 Sanitary Measures in Rearing Facilities

Numerous manuals for biosafety practices in laboratory facilities exist (Liberman 1995; Richmond and McKinney 1998; World Health Organization 2004; Diane and Hunt 2006; Furr 2000). Although pertinent, they are as a rule very extensive

(200–800 pages) providing an exhaustive description of procedures and protocols for laboratory biosafety. We provide a summarized description of procedures required in laboratory facilities in general, and when dealing with pathogen symptomatic insect reared colonies, in particular.

The following procedures should be followed for each individual work session when performing methodologies mentioned in this handbook, to be performed in a fume hood cabinet:

When possible, fume hoods should be on at all time filtering air inside the cabinet. If not possible, the purge air flow should be activated at least 15 min prior to use.

1. Surface disinfection should be carried out for each session. Ethyl or isopropyl alcohol at 70–90% concentration is a general disinfectant in laboratory facilities for primary disinfection. Ultra violet germicidal lamps should be also incorporated to the fume hood system for weekly surface sterilization of the fume hood. Laboratory personnel should never be exposed to UV irradiation.
2. All materials required to complete a session (sterilized glassware, scalpels, pincers, probes or other equipment) are placed inside the cabinet prior to work and opening sealed containers.
3. Sterilization of hands with alcohol and use of latex gloves prior do initiate fume hood session is also required by laboratory personnel. Laboratory coat sleeves should not cover forearms, which should be also disinfected or covered with long gloves.
4. Work is performed at least four inches forward to the glass panel in the front of the hood. Disinfection during the session of metallic utensils used to contact the microorganisms should be made used a Bunsen burner located outside the fume hood.

Upon completion of al procedures inside the fume hood, all disposable materials are separated, and sealed in proper container packages (e.g. Fisherbrand autoclave bag with universal label for potentially infectious material) and subsequently autoclaved; Utensils used in the fume hood session should be also sterilized with heat or UV light equipment.

The hood should be left on for at least five minutes to purge the air.

Personal isolation equipment and/or a barrier gown are usually used to access safety quarantine zones and should be disposed and/or sterilized after leaving the facilities. Latex gloves are highly recommended in quarantine facilities. Directional airflow into the room and an exhaust air system from the room should be installed. For aerosol microorganisms it is imperative to use safety respirator masks approved by the National Institute for Occupational Safety and Health (NIOSH). Respirator masks can range from disposable respirators used to filter nontoxic dusts or spray particles to full face respiratory masks with a two filter system for toxic microorganisms.

Depending on the organism, decontamination procedures when leaving quarantine rooms also are required. Access to this type of facilities should ensure that only authorized personnel may be admitted. Usually, a specific quarantine facility operations manual is adopted depending on the type or organisms to be handled and its hazardous level.

All waste materials from experiments with potentially infectious materials should be sterilized. Autoclaves are commonly used in laboratory facilities to dispose of culture dishes, pipette tips and autoclavable glassware, contaminated paper and tissues, etc., as well as living cultures and culture liquors. Autoclaved glassware may be then returned to the normal wash-up procedures. Sterilization using autoclave equipment requires approx. 15 psi pressure (1 atmosphere) with a chamber temperature of at least 250 F (121 C). Time length for sterilization depends on the volume of the load, ranging from 30 to 60 min. Waste which has been autoclaved can be disposed with regular waste providing that biohazard labeling in the waste bags is legible after sterilization.

If spills occur, the following procedures should be put into action:

1. Remove any contaminated clothing and wash exposed skin with disinfectant.
2. Cover the spill with paper towels, pour concentrated disinfectant around spill allowing it to mix with the spill. Allow at least 20 min contact time.
3. Wait at least 30 min before reentering the laboratory to allow dissipation of aerosols created by the spill.
4. Use protective clothing (long sleeved gown, gloves, and shoe covers) to recover spill material and containers. Depending on the nature of the spill, it may be advisable to wear a respirator with High Efficiency Particulate Air (HEPA) cartridges.
5. Use forceps to place sharp objects into a sharps container. Wipe surrounding surfaces with disinfectant to cover all splash areas.
6. Place all contaminated materials, including protective clothing, into a biohazard bag and autoclave.

References

Diane O, Hunt DL (2006) Biological safety: principles and practices, 4th edn. ASM Press, Virginia, USA

Furr AK (2000) CRC handbook of laboratory safety, 5th edn. CRC Press, Boca Raton, FL

Gouli VV, Parker BL, Reid W (1999) Method for rearing gnotobiotic thrips. J Appl Ent 123: 127–128

Liberman D (ed) (1995) Biohazards management handbook. Marcel Dekker Inc., New York

Richmond JY, McKinney RW (1998) Biosafety in microbiological and biomedical laboratories. Centers for disease control. Atlanta, GA, and National Institutes of Health, Bethesda, MD. HHS Publication number (CDC) 93–8395

Talpalatsky PL, Mikhaelev AI, Gouli VV, Ivanov GM, Sukharukova AN (1984) Gnotobiotic insects as a model to study experimental pathology. Collected articles of the all-Union Ins. Biol. Pl. Protection: Mass-rearing of insects, All-Union Institute Kishinev, Moldova 5–16 (In Russian, summary in English)

World Health Organization (2004) Laboratory biosafety manual, 3rd edn. World Health Organization Press, Beijing